T0076690

Reviews of Environmental Contamination and Toxicology

VOLUME 216

For further volumes:
http://www.springer.com/series/398

Aquatic Life Water Quality Criteria for Selected Pesticides

Editor
Ronald S. Tjeerdema

VOLUME 216

Editor
Ronald S. Tjeerdema
Department of Environmental Toxicology
College of Agricultural and Environmental Sciences
University of California
Davis, CA, USA

Please note that additional material for this book can be downloaded from
http://extras.springer.com

ISSN 0179-5953
ISBN 978-1-4614-2259-4 e-ISBN 978-1-4614-2260-0
DOI 10.1007/978-1-4614-2260-0
Springer New York Dordrecht Heidelberg London

Library of Congress Control Number: 2011945517

Printed on acid-free paper

Springer is part of Springer Science+Business Media (www.springer.com)

Special Foreword

California's Central Valley has been a leader in agricultural productivity since it was first settled by European immigrants in the nineteenth century. Drained by both the Sacramento and San Joaquin River Systems, its fertile farmlands represent the most productive region in the USA today. Key to that productivity is the use of modern agrochemicals, including fertilizers and pest control agents. However, while enormously useful as tools, they also present their share of risks to both human health and the environment.

The Central Valley also contains a rich, endemic flora and fauna—both terrestrial and aquatic. Thus, the challenge for many years has been how to enhance agricultural productivity in the region while maintaining environmental quality, as agricultural residues pose a risk to not only the valley, but also the San Francisco Bay-Delta Region. In recent years, the California State Water Resources Control Board, through its Regional Water Quality Control Board (RWQCB; Central Valley Region), has sought to better characterize the risk to endemic aquatic organisms posed by agricultural pesticides used in the valley. Such characterization would assist in guiding the continued use of pesticides in an environmentally safe manner. However, methods for assessing the risk of pesticides to aquatic species have been slow to develop. Therefore, the RWQCB approached us a number of years ago with the request that we develop an advanced method for assessing such risk, and then apply it to develop criteria for the continued safe use of many of the most effective agents available today.

In response, we first surveyed the methods currently available worldwide—published in *Reviews of Environmental Contamination and Toxicology* (Volume 199). We subsequently developed an advanced method (the University of California—Davis Methodology)—also published in *Reviews* (Volume 209)—which significantly built upon the early progress of others. We then applied the methodology to develop risk criteria for representative agents from three pesticide classes: organophosphates, pyrethroids, and substituted ureas. Those papers are the subject of this volume—providing guidance on the safe use of pesticides, not only in California's Central Valley but potentially worldwide. It should be noted that the assessment of risk is not a one-time task but an ongoing process, as criteria can be continually refined by the addition of new, and potentially high-quality, data to decrease uncertainty in the derived values over time. In fact, to our knowledge water quality criteria for the pyrethroids have not been previously derived in the USA. Thus, the wide-ranging review of each chemical presented in the subsequent papers represents a good foundation for future refinements.

Another useful aspect of the risk assessment process is that data gaps can be identified—which may stimulate new research to fill them. For instance, there is currently a lack of chronic toxicity data for all seven targeted pesticides (chlorpyrifos, diazinon, malathion, bifenthrin, cyfluthrin, cypermethrin, lambda-cyhalothrin, permethrin and diuron), and because of this the uncertainty of the derived chronic criteria could not be quantified. High-quality tests using flow-through exposure systems which generate calculated toxicity values based on measured concentrations are needed for all the agents but particularly the pyrethroids, which are highly sorptive. The influence of both temperature and nonadditive mixture effects also need further documentation so that they may be incorporated into criteria compliance.

The authors of the papers presented in this volume (Tessa Fojut, Amanda Palumbo, Patti TenBrook, and Isabel Faria) possess a wealth of experience in toxicology and environmental chemistry—as well as environmental risk assessment. It is through their tireless efforts that these criteria are now available with the hope that their application will facilitate the continued use of the subject agents in an environment-friendly manner. I am particularly grateful to Tessa Fojut for her many efforts in preparing the final criteria manuscripts for publication.

Davis, CA, USA Ronald S. Tjeerdema

Foreword

International concern in scientific, industrial, and governmental communities over traces of xenobiotics in foods and in both abiotic and biotic environments has justified the present triumvirate of specialized publications in this field: comprehensive reviews, rapidly published research papers and progress reports, and archival documentations. These three international publications are integrated and scheduled to provide the coherency essential for nonduplicative and current progress in a field as dynamic and complex as environmental contamination and toxicology. This series is reserved exclusively for the diversified literature on "toxic" chemicals in our food, our feeds, our homes, recreational and working surroundings, our domestic animals, our wildlife, and ourselves. Tremendous efforts worldwide have been mobilized to evaluate the nature, presence, magnitude, fate, and toxicology of the chemicals loosed upon the Earth. Among the sequelae of this broad new emphasis is an undeniable need for an articulated set of authoritative publications, where one can find the latest important world literature produced by these emerging areas of science together with documentation of pertinent ancillary legislation.

Research directors and legislative or administrative advisers do not have the time to scan the escalating number of technical publications that may contain articles important to current responsibility. Rather, these individuals need the background provided by detailed reviews and the assurance that the latest information is made available to them, all with minimal literature searching. Similarly, the scientist assigned or attracted to a new problem is required to glean all literature pertinent to the task, to publish new developments or important new experimental details quickly, to inform others of findings that might alter their own efforts, and eventually to publish all his/her supporting data and conclusions for archival purposes. In the fields of environmental contamination and toxicology, the sum of these concerns and responsibilities is decisively addressed by the uniform, encompassing, and timely publication format of the Springer triumvirate:

Reviews of Environmental Contamination and Toxicology [Vol. 1 through 97 (1962–1986) as Residue Reviews] for detailed review articles concerned with any aspects of chemical contaminants, including pesticides, in the total environment with toxicological considerations and consequences.

Bulletin of Environmental Contamination and Toxicology (Vol. 1 in 1966) for rapid publication of short reports of significant advances and discoveries in the fields of air, soil, water, and food contamination and pollution as well as methodology and other disciplines concerned with the introduction, presence, and effects of toxicants in the total environment.

Archives of Environmental Contamination and Toxicology (Vol. 1 in 1973) for important complete articles emphasizing and describing original experimental or theoretical research work pertaining to the scientific aspects of chemical contaminants in the environment.

Manuscripts for Reviews and the Archives are in identical formats and are peer reviewed by scientists in the field for adequacy and value; manuscripts for the Bulletin are also reviewed, but are published by photo-offset from camera-ready copy to provide the latest results with minimum delay. The individual editors of these three publications comprise the joint Coordinating Board of Editors with referral within the board of manuscripts submitted to one publication but deemed by major emphasis or length more suitable for one of the others.

Coordinating Board of Editors

Preface

The role of *Reviews* is to publish detailed scientific review articles on all aspects of environmental contamination and associated toxicological consequences. Such articles facilitate the often complex task of accessing and interpreting cogent scientific data within the confines of one or more closely related research fields.

In the nearly 50 years since *Reviews of Environmental Contamination and Toxicology (formerly Residue Reviews)* was first published, the number, scope, and complexity of environmental pollution incidents have grown unabated. During this entire period, the emphasis has been on publishing articles that address the presence and toxicity of environmental contaminants. New research is published each year on a myriad of environmental pollution issues facing people worldwide. This fact, and the routine discovery and reporting of new environmental contamination cases, creates an increasingly important function for *Reviews*.

The staggering volume of scientific literature demands remedy by which data can be synthesized and made available to readers in an abridged form. Reviews addresses this need and provides detailed reviews worldwide to key scientists and science or policy administrators, whether employed by government, universities, or the private sector.

There is a panoply of environmental issues and concerns on which many scientists have focused their research in past years. The scope of this list is quite broad, encompassing environmental events globally that affect marine and terrestrial ecosystems; biotic and abiotic environments; impacts on plants, humans, and wildlife; and pollutants, both chemical and radioactive; as well as the ravages of environmental disease in virtually all environmental media (soil, water, air). New or enhanced safety and environmental concerns have emerged in the last decade to be added to incidents covered by the media, studied by scientists, and addressed by governmental and private institutions. Among these are events so striking that they are creating a paradigm shift. Two in particular are at the center of everincreasing media as well as scientific attention: bioterrorism and global warming. Unfortunately, these very worrisome issues are now superimposed on the already extensive list of ongoing environmental challenges.

The ultimate role of publishing scientific research is to enhance understanding of the environment in ways that allow the public to be better informed. The term "informed public" as used by Thomas Jefferson in the age of enlightenment conveyed the thought of soundness and good judgment. In the modern sense, being "well informed" has the narrower meaning of having access to sufficient information. Because the public still gets most of its information on science and technology from TV news and reports, the role for scientists as interpreters and brokers of scientific information to the public will grow rather than diminish. Environmentalism is the newest global political force, resulting in the emergence of multinational consortia to control pollution and the evolution of the environmental ethic. Will the new politics of the twenty-first century involve a consortium of technologists and environmentalists, or a progressive confrontation? These matters are of genuine concern to governmental agencies and legislative bodies around the world.

For those who make the decisions about how our planet is managed, there is an ongoing need for continual surveillance and intelligent controls to avoid endangering the environment, public health, and wildlife. Ensuring safety-in-use of the many chemicals involved in our highly industrialized culture is a dynamic challenge, for the old, established materials are continually being displaced by newly developed molecules more acceptable to federal and state regulatory agencies, public health officials, and environmentalists.

Reviews publishes synoptic articles designed to treat the presence, fate, and, if possible, the safety of xenobiotics in any segment of the environment. These reviews can be either general or specific, but properly lie in the domains of analytical chemistry and its methodology, biochemistry, human and animal medicine, legislation, pharmacology, physiology, toxicology, and regulation. Certain affairs in food technology concerned specifically with pesticide and other food-additive problems may also be appropriate.

Because manuscripts are published in the order in which they are received in final form, it may seem that some important aspects have been neglected at times. However, these apparent omissions are recognized, and pertinent manuscripts are likely in preparation or planned. The field is so very large and the interests in it are so varied that the editor and the editorial board earnestly solicit authors and suggestions of underrepresented topics to make this international book series yet more useful and worthwhile.

Justification for the preparation of any review for this book series is that it deals with some aspect of the many real problems arising from the presence of foreign chemicals in our surroundings. Thus, manuscripts may encompass case studies from any country. Food additives, including pesticides, or their metabolites that may persist into human food and animal feeds are within this scope. Additionally, chemical contamination in any manner of air, water, soil, or plant or animal life is within these objectives and their purview.

Manuscripts are often contributed by invitation. However, nominations for new topics or topics in areas that are rapidly advancing are welcome. Preliminary communication with the editor is recommended before volunteered review manuscripts are submitted.

Summerfield, NC David M. Whitacre

Contents

Aquatic Life Water Quality Criteria Derived via the UC Davis Method: I. Organophosphate Insecticides

Amanda J. Palumbo, Patti L. TenBrook, Tessa L. Fojut, Isabel R. Faria, and Ronald S. Tjeerdema

1 Introduction

Water quality criteria are numeric concentrations for chemicals in water bodies that, if not exceeded, should protect aquatic wildlife from toxic effects of those chemicals. These criteria, which do not consider economics or societal values, typically are derived using the existing toxicity data. Water quality criteria can be used as a basis to set legal and enforceable water quality standards or objectives in accordance with the Clean Water Act.

A new methodology for deriving freshwater pesticide water quality criteria for the protection of aquatic life was developed by the University of California Davis (TenBrook et al. 2010). The need for a new methodology was identified by a review of existing methodologies (TenBrook et al. 2009) that was commissioned by the California Central Valley Regional Water Quality Control Board (CVRWQCB). New research in the fields of aquatic toxicology and risk assessment has been incorporated into the UC Davis methodology (UCDM), whereas the United States Environmental Protection Agency (USEPA) method for derivation of aquatic life criteria has not been updated since 1985 (USEPA 1985). The fundamentals of the

A.J. Palumbo • T.L. Fojut (✉) • I.R. Faria • R.S. Tjeerdema
Department of Environmental Toxicology, College of Agricultural and Environmental Sciences, University of California, Davis, CA 95616-8588, USA
e-mail: tlfojut@ucdavis.edu

P.L. TenBrook
Department of Environmental Toxicology, College of Agricultural and Environmental Sciences, University of California, Davis, CA 95616-8588, USA

Current affiliation: USEPA Region 9*, 75 Hawthorne Street, San Francisco, CA 94195, USA
*Contents do not necessarily reflect the views or policies of the USEPA nor does mention of trade names or commercial products constitute endorsement or recommendation for use.

R.S. Tjeerdema (ed.), *Aquatic Life Water Quality Criteria for Selected Pesticides*, Reviews of Environmental Contamination and Toxicology 216,
DOI 10.1007/978-1-4614-2260-0_1,

new method are similar to those of the USEPA (1985) approach, in that a species sensitivity distribution (SSD) is the preferred method of criteria calculation and an acute-to-chronic ratio (ACR) is used when chronic data are limited. Some of the major differences provided by the UCDM are a thorough and transparent study evaluation procedure; a more advanced SSD; alternate procedures if data requirements for the SSD or ACR cannot be met; and inclusion of mixtures.

The UCDM has been used to derive aquatic life criteria for several pesticides of particular concern in the Sacramento River and San Joaquin River watersheds, which are also widely used throughout the USA. This paper is the first in a series in which criteria were derived for three organophosphate (OP) insecticides (chlorpyrifos, diazinon, and malathion), five pyrethroid insecticides (bifenthrin, cyfluthrin, cypermethrin, lambda-cyhalothrin, and permethrin), and one phenyl-urea herbicide (diuron). Diazinon and chlorpyrifos were chosen as the first pesticides to be evaluated with the UCDM because there were already national and state criteria for these compounds to which the results of the UCDM could be compared; malathion was included in the analysis because it is another organophosphate pesticide that is of concern for water quality. The UCDM contains detailed procedures for criteria derivation, as well as the rationale behind the selection of specific methods (TenBrook et al. 2010). This organophosphate criteria derivation article describes the procedures used to derive criteria according to the UCDM, and provides several references to specific sections numbers of the UCDM document (TenBrook et al. 2010) so that the reader may refer to the UCDM for further details.

2 Data Collection and Evaluation

Chlorpyrifos (*O*,*O*-diethyl *O*-(3,5,6-trichloro-2-pyridinyl) phosphorothioate), diazinon (*O*,*O*-diethyl *O*-2-isopropyl-6-methylpyrimidin-4-yl phosphorothioate), and malathion (diethyl 2-dimethoxyphosphinothioylsulfanylbutanedioate) are organophosphate insecticides. The physical–chemical properties of these OPs (Table 1) indicate that some fraction remains dissolved in the water column and eventually degrades there (Table 2), some fraction partitions to the sediments, and that they are not likely to volatilize from the water column.

Original studies on the effects of chlorpyrifos (~340), diazinon (~250), and malathion (~200) on aquatic life were identified and reviewed. Studies were from both the open literature and unpublished studies submitted to the USEPA and California Department of Pesticide Regulation (CDPR) by pesticide registrants. Unpublished studies held by these agencies can be requested from the respective agencies; the full request instructions to acquire them are given in the UCDM (TenBrook et al. 2010). To determine the usefulness of these studies for criteria derivation, they were subjected to a review process, depending on the type of study; the three types were (1) single-species effects, (2) ecosystem-level studies, and (3) terrestrial wildlife studies.

Table 1 Summary of physical–chemical properties

	Chlorpyrifos	Diazinon	Malathion
Molecular weight	350.6	304.36	330.358
Density (g/mL)	1.44[c] (20°C)	1.11[j] (20°C)	1.22 (geomean, $n = 3$)[o]
Water solubility (mg/L)	1.46 (geomean, $n = 3$)[d]	46.0 (geomean, $n = 4$)[k]	146.16 (geomean, $n = 7$)[p]
Melting point (°C)	42.73 (geomean of extremes)[e]	Liquid at room temperature[c]	2.43 (geomean, $n = 4$)[q]
Vapor pressure (Pa)	2.36×10^{-3} (geomean, $n = 3$)[f]	0.014 (geomean, $n = 3$)[c,l]	1.20×10^{-3} (geomean, $n = 6$)[r]
Henry's law constant (K_H) (Pa m^3/mol)	0.640 (geomean, $n = 4$)[c,g]	0.0114 (geomean, $n = 2$)[m]	1.65×10^{-3} (geomean, $n = 4$)[s]
Log K_{oc}[a]	4.06 (geomean, $n = 2$)[h]	3.35 (geomean, $n = 8$)[n]	2.77 (geomean, $n = 8$)[t]
Log K_{ow}[b]	4.96[i]	3.81[i]	2.84 (geomean, $n = 8$)[c,u]

[a] Log K_{oc}: log-normalized organic carbon–water partition coefficient
[b] Log K_{ow}: log-normalized octanol–water partition coefficient
[c] Tomlin (2003)
[d] Hummel and Crummet (1964), Felsot and Dahm (1979), and Drummond (1986)
[e] Bowman and Sans (1983a), Brust (1964, 1966), McDonald et al. (1985), and Rigertink and Kenaga (1966)
[f] Brust (1964), McDonald et al. (1985), and Chakrabarti and Gennrich (1987)
[g] Wu et al. (2002), Fendinger and Glotfelty (1990), and Downey (1987)
[h] Racke (1993) and Spieszalski et al. (1994)
[i] Sangster Research Laboratories (2004)
[j] Worthing (1991)
[k] Martin and Worthing (1977), Jarvinen and Tanner (1982), Kanazawa (1981), and Bowman and Sans (1979, 1983b)
[l] Kim et al. (1984) and Hinckley et al. (1990)
[m] Fendinger and Glotfelty (1988) and Fendinger et al. (1989)
[n] Iglesias-Jimenez et al. (1997), Cooke et al. (2004), and Kanazawa (1989)
[o] Barton (1988), Mackay et al. (2006), and Verschueren (1996)
[p] Kidd and James (1991), Howard (1989), Cheminova (1988), Kamrin and Montgomery (2000), Kabler (1989), and Hartley and Graham-Bryce (1980)
[q] Lide (2004), Kidd and James (1991), Budavari et al. (1996), Howard (1989), and Barton (1988)
[r] Howard (1989), Hartley and Graham-Bryce (1980), Verschueren (1996), Kidd and James (1991), Tondreau (1987), Melnikov (1971), and Barton (1988)
[s] Howard (1989), Mackay et al. (2006), and Kamrin and Montgomery (2000)
[t] Mackay et al. (2006), Kamrin and Montgomery (2000), Karickhoff (1981), and Sabljic et al. (1995)
[u] Kamrin and Montgomery (2000), Howard (1989), Barton (1988), Verschueren (1996), and Mackay et al. (2006)

Table 2 Environmental fate of chlorpyrifos, diazinon, and malathion

	Chlorpyrifos	Diazinon	Malathion
Hydrolysis half-life (days)	210 (pH 4.7/15°C)[a]	0.49 (pH 3.1/20°C)[f]	40 (pH 8/0°C)[m]
	99.0 (pH 6.9/15°C)[a]	6 (pH 10.4/20°C)[f]	36 h (pH 8/27°C)[m]
	54.2 (pH 8.1/15°C)[a]	17 (pH 8.0/40°C)[g]	1 h (pH 8/40°C)[m]
	120 (pH 6.1/20°C)[b]	30 (pH 7.4–7.8/22.5°C)[h]	10.5 (pH 7.4/20°C)[n]
	53 (pH 7.4/20°C)[b]	31 (pH 5.0/20°C)[f]	1.3 (pH 7.4/37.5°C)[n]
	62.7 (pH 4.7/25°C)[a]	37.2[i]	107 (pH 5/25°C)[o]
	77 (pH 5.9/25°C)[c]	52 (pH 7.3/22°C)[j]	6.21 (pH 7, 25°C)[o]
	204 (pH 6.1/25°C)[c]	69 (pH 6.1/22°C)[j]	0.49 (pH 9, 25°C)[o]
	35.3 (pH 6.9/25°C)[a]	80 (pH 7.3/22°C)[j]	
	22.8 (pH 8.1/25°C)[a]	88 (pH 8.0/24°C)[h]	
	15 (pH 9.7/25°C)[c]	136 (pH 9.0/20°C)[f]	
	15.7 (pH 4.7/35°C)[a]	171 (pH 7.3/21°C)[k]	
	11.5 (pH 6.9/35°C)[a]	185 (pH 7.4/20°C)[f]	
	4.5 (pH 8.1/35°C)[a]		
Aqueous photolysis (days)	13.9 (pH 5.0)[d]	9–12 (25°C)[l]	156 (pH 4/25°C)[p]
	21.7(pH 6.9)[d]		94 (pH 4, 25°C)[p]
	13.1(pH 8.0)[d]		
	31(pH 7.0)[e]		
	43 (pH 7.0)[e]		
	345 (pH 7.0)[e]		

NR not reported
[a] Meikle and Youngson (1978)
[b] Freed et al. (1979a)
[c] Macalady and Wolfe (1983)
[d] Meikle et al. (1983)
[e] Dilling et al. (1984)
[f] Gomaa et al. (1969) and Faust and Gomaa (1972)
[g] Noblet et al. (1996)
[h] Jarvinen and Tanner (1982)
[i] Medina et al. (1999)
[j] Lartiges and Garrigues (1995)
[k] Mansour et al. (1999)
[l] Kamiya and Kameyama (1998)
[m] Wolfe et al. (1977)
[n] Freed et al. (1979b)
[o] Teeter (1988)
[p] Carpenter (1990)

Single-species effects studies were evaluated in a two-step numeric scoring process. First, studies were evaluated based on six main criteria: (1) use of a control; (2) freshwater species; (3) species belongs to a family in North America; (4) chemical purity >80%; (5) end point linked to survival, growth, or reproduction; and (6) a toxicity value was calculated or is calculable. Studies that met all of these parameters were rated relevant (R) while studies that did not meet one or two of the six relevance criteria were rated less relevant (L). Finally, studies that lacked more than two of these criteria were considered to be not relevant (N). The studies rated as relevant (R) or less relevant (L) were subject to a second evaluation while those that rated as not relevant (N) were not considered further. Data summaries detailing study parameters and scoring for all studies are included as the Supporting Material (http://extras.springer.com/).

The second review of the studies rated R or L was designed to evaluate data reliability. Reliability scores were based on if test parameters were reported and the acceptability of those parameters according to standard methods; some of the scored test parameters were organism source and care, control description and response, chemical purity, concentrations tested, water quality conditions, and statistical methods. Numeric scores were translated into ratings of reliable (R), less reliable (L), or not reliable (N). Each study was given a two-letter code, with the first letter corresponding to the relevance rating and the second letter corresponding to the reliability rating. Acceptable studies, rated as relevant and reliable (RR), were used for numeric criteria derivation. Supplemental studies, rated as relevant and less reliable (RL), less relevant and reliable (LR) or less relevant and less reliable (LL), were not used directly for criteria calculation, but were used for evaluation of the criteria to check that they are protective of particularly sensitive species and threatened and endangered species, which may not be represented in the RR data sets. Data that were rated as acceptable (RR) for criteria derivation are summarized in Tables 3–8. All other toxicity data are available as the Supporting Material (http://extras.springer.com/). Studies that were rated not relevant (N) or relevant or less relevant, but not reliable (RN or LN), were not used in any aspect of criteria derivation.

Mesocosm, microcosm, and ecosystem (field and laboratory) studies were subject to a separate evaluation of reliability. Studies that were rated reliable (R) or less reliable (L) were used to evaluate the derived criteria to ensure that they are protective of ecosystems. Terrestrial wildlife toxicity studies for mallard ducks were evaluated specifically for the consideration of bioaccumulation. Mallard duck studies that were rated reliable (R) or less reliable (L) were used in estimations of bioaccumulative potential.

3 Data Reduction

Multiple toxicity values for each pesticide for the same species were combined into one species mean acute value (SMAV) or one species mean chronic value (SMCV) by calculating the geometric mean of appropriate values. To arrive at one SMAV or SMCV per species, some data rated RR were excluded from the final RR data set for the following reasons: tests that used measured concentrations are preferred over tests that used nominal concentrations; flow-through tests are preferred over static tests; a test with a more sensitive life stage of the same species was available; longer exposure durations are preferred; tests at standard conditions are preferred over those at nonstandard conditions; and tests with more sensitive end points were available. Acceptable acute and chronic data that were excluded, and the reasons for their exclusion, are shown in Tables S1–S3 (Supporting Material http://extras.springer.com/). For chlorpyrifos, the final acceptable data sets contain 17 SMAVs

Table 3 Final acute toxicity data set for chlorpyrifos

Species	Test type	Meas/Nom	Chemical grade (%)	Duration (h)	Temp (°C)	End point	Age/size	LC/EC_{50} (μg/L)	References
Ceriodaphnia dubia	S	Meas	99.0	96	25	Mortality	<24 h	0.053	Bailey et al. (1997)
C. dubia	S	Meas	99.0	96	25	Mortality	<24 h	0.055	Bailey et al. (1997)
C. dubia	SR	Meas	99.0	96	24.6	Mortality	<24 h	0.13	CDFG (1992e)
C. dubia	SR	Meas	99.0	96	24.3	Mortality	<24 h	0.08	CDFG (1992b)
C. dubia	SR	Meas	99.8	96	24.6	Survival	<24 h	0.0396	CDFG (1999)
Geometric mean								0.0654	
Chironomus tentans	S	Meas	98.0	96	21	Immobility	Third to fourth instar	0.16	Belden and Lydy (2006)
C. tentans	S	Meas	90.0	96	21	Immobility	Fourth instar	0.17	Lydy and Austin (2005)
C. tentans	S	Meas	98.0	96	20	Immobility + mortality	Fourth instar	0.39	Belden and Lydy (2000)
Geometric mean								0.220	
Daphnia ambigua	S	Meas	99.0	48	21	Immobility	Neonates	0.035	Harmon et al. (2003)
Daphnia magna	S	Meas	99.0	48	19.5	Mortality	<24 h	1.0	Kersting and Van Wijngaarden (1992)
D. magna	FT	Nom (most)	95.5	48	18–21	Mortality	<24 h	0.10	Burgess (1988)
Geometric mean								0.32	
Daphnia pulex	S	Meas	Technical	48	20	Immobility	<24 h	0.25	Van Der Hoeven and Gerritsen (1997)
Hyalella azteca	S	Meas	90.0	96	20	Mortality	14–21 days	0.0427	Anderson and Lydy (2002)
H. azteca	SR	Meas	98.1	96	19	Mortality	14–21 days	0.138	Brown et al. (1997)
Geometric mean								0.077	
Ictalurus punctatus	FT	Meas	99.9	96	17.3	Mortality	7.9 g	806	Phipps and Holcombe (1985)

Lepomis macrochirus	FT	Meas	99.9	96	17.3	Mortality	0.8 g	10	Phipps and Holcombe (1985)
L. macrochirus	FT	Meas	99.9	96	22	Mortality	2.1 g	5.8	Bowman (1988)
Geometric mean								7.6	
Neomysis mercedis	SR	Meas	99.0	96	17.4	Mortality	<5 days	0.15	CDFG (1992d)
N. mercedis	SR	Meas	99.0	96	17.2	Mortality	<5 days	0.16	CDFG (1992a)
N. mercedis	SR	Meas	99.0	96	17.1	Mortality	<5 days	0.14	CDFG (1992c)
Geometric mean								0.150	
Oncorhynchus mykiss	FT	Meas	99.9	96	12	Mortality	Juvenile	8.0	Holcombe et al. (1982)
O. mykiss	FT	Meas	95.9	96	12	Mortality	0.25 g	25.0	Bowman (1988)
Geometric mean								14	
Oncorhynchus tshawytscha	SR	Meas	99.5	96	14.8	Mortality	Juvenile	15.96	Wheelock et al. (2005)
Orconectes immunis	FT	Meas	99.9	96	17.3	Mortality	1.8 g	6	Phipps and Holcombe (1985)
Pimephales promelas	FT	Meas	99.9	96	25	Mortality	32 days	200	Geiger et al. (1988)
P. promelas	FT	Meas	99.9	96	25	Mortality	31–32 days	203	Holcombe et al. (1982)
P. promelas	FT	Meas	98.7	96	25	Mortality	Newly hatched	140	Jarvinen and Tanner (1982)
Geometric mean								178	
Procloeon sp.	SR	Meas	99	48	21.3	Mortality	0.5–1.0 cm	0.1791	Anderson et al. (2006)
Procloeon sp.	SR	Meas	99	48	21.3	Mortality	0.5–1.0 cm	0.0704	Anderson et al. (2006)
Procloeon sp.	SR	Meas	99	48	21.3	Mortality	0.5–1.0 cm	0.0798	Anderson et al. (2006)

(continued)

Table 3 (continued)

Species	Test type	Meas/Nom	Chemical grade (%)	Duration (h)	Temp (°C)	End point	Age/size	LC/EC$_{50}$ (μg/L)	References
Geometric mean								0.100	
Pungitius pungitius	FT	Meas	99.8	96	19	Mortality	Adult	4.7	Van Wijngaarden et al. (1993)
Simulium vittatum IS-7	S	Meas	98.0	24	19	Mortality	Second and third instar	0.06	Hyder et al. (2004)
Xenopus laevis	SR	Nom	99.80	96	24.7	Mortality	<24 h	2,410	El-Merhibi et al. (2004)

All studies were rated relevant and reliable (RR) and were conducted at standard temperature (Standard temperatures are particular for each species. See standard methods referenced in Tables 9 and 10 of TenBrook et al. (2010))

S static, *SR* static renewal, *FT* flow through

Table 4 Final chronic toxicity data set for chlorpyrifos

Species	Test type	Meas/Nom	Chemical grade (%)	Duration (days)	Temp (°C)	End point	Age/size	NOEC (μg/L)	LOEC (μg/L)	MATC (μg/L)	Reference
Ceriodaphnia dubia	SR	Meas	99.8	7	24.6	Mortality	<24 h	0.029	0.054	0.0396	CDFG (1999)
C. dubia	SR	Meas	99.8	7	24.6	Reproduction	<24 h	0.029	0.054	0.0396	CDFG (1999)
Geometric mean								0.029	0.054	0.0396	
Pimephales promelas	FT	Meas	98.7	60	24.3–25.9	Growth	<24 h	0.63	1.21	0.87	Jarvinen et al. (1983)
P. promelas	FT	Meas	98.7	32	23.5–26.0	Weight	Newly hatched	1.6	3.2	2.3	Jarvinen and Tanner (1982)
P. promelas	FT	Meas	99.7	25 and 32	25.0–25.5	F_0 and F_1 Mortality	<24 h	0.568	1.093	0.788	Mayes et al. (1993)
Geometric mean								0.83	1.62	1.16	
Neomysis mercedis	SR	Meas	99.0	96	17	Mortality	<5 days	0.001[a]			CDFG (1992a)
N. mercedis	SR	Meas	99.0	96	17	Mortality	<5 days	0.001[a]			CDFG (1992d)
Geometric mean								0.001[a]			

All studies were rated relevant and reliable (RR) and were conducted at standard temperatures for a given species
SR static renewal, *FT* flow through
[a] Chronic values for *Neomysis mercedis* were estimated from acute data

Table 5 Final acute toxicity data set for diazinon

Species	Test type	Meas/ Nom	Chemical grade (%)	Duration (h)	Temp (°C)	End point	Age/size	LC/EC_{50} (μg/L)	Reference
Ceriodaphnia dubia	SR	Meas	87.3	96	24.7	Mortality	<24 h	0.436 (0.342–0.504)	CDFG (1998a)
C. dubia	SR	Meas	88.0	96	24.4	Mortality	<24 h	0.47	CDFG (1992f)
C. dubia	SR	Meas	88.0	96	24.4	Mortality	<24 h	0.507 (0.42–0.71)	CDFG (1992g)
C. dubia	S	Meas	99.0	96	25	Mortality	<24 h	Test 1: 0.32 (0.27–0.38) Test 2: 0.35 (0.32–0.38)	Bailey et al. (1997)
C. dubia	S	Meas	99.0	48	25	Mortality	<24 h	Test 3: 0.26 (0.21–0.32) Test 4: 0.29 (0.19–0.46)	Bailey et al. (1997)
C. dubia	S	Meas	Analytical	48	25	Mortality	<24 h	0.33	Bailey et al. (2000)
C. dubia	S	Meas	99.0	48	25	Mortality	<24 h	Test 1: 0.38 Test 2: 0.33	Bailey et al. (2001)
C. dubia	S	Meas	99.8	48	25	Mortality	<24 h	0.21	Banks et al. (2005)
Geometric mean								0.34	
Chironomus dilutus (formerly *tentans*)	S	Nom	95.0	96	23	Mortality/ immobility	Third instar	10.7 (7.55–15.2)	Ankley and Collyard (1995)
Daphnia magna	FT	Meas	87.7	96	20	Mortality/ immobility	<24 h	0.52 (0.32–0.83)	Surprenant (1988)
Gammarus pseudolimnaeus	S/R	Meas	100.0	96	18	Mortality	Mature	16.82 (12.82–22.08)	Hall and Anderson (2005)
Hyalella azteca	S	Meas	98.0	96	20	Mortality	14–21 days	4.3 (3.7–5.6)	Anderson and Lydy (2002)

Jordanella floridae	FT	Meas	92.5	96	25	Mortality	6–7 weeks	Test 1: 1,500 (1,200–1,900) Test 2: 1,800 (1,600–2,000)	Allison and Hermanutz (1977)
Geometric mean								1,643	
Lepomis macrochirus	FT	Meas	92.5	96	25	Mortality	1 year	Test 1: 480 (340–670) Test 2: 440 (310–620)	Allison and Hermanutz (1977)
Geometric mean								460	
Neomysis mercedis	S/R	Meas	88.0	96	17	Mortality	<5 days	3.57 (2.99–4.36)	CDFG (1992h)
N. mercedis	S/R	Meas	88.0	96	17.5	Mortality	<5 days	4.82 (3.95–6.00)	CDFG (1992i)
Geometric mean								4.15	
Physa spp.	S/R	Meas	87.0	96	21.6	Mortality	Juvenile	4,441	CDFG (1998b)
Pimephales promelas	FT	Meas	92.5	96	25	Mortality	15–20 weeks	Test 1: 6,800 Test 2: 6,600 Test 3: 10,000	Allison and Hermanutz (1977)
P. promelas	FT	Meas	87.1	96	24.5	Mortality	31 days	9,350 (8,120–10,800)	Geiger et al. (1988)
P. promelas	FT	Meas	87.1	96	23.5–26	Mortality	Newly hatched	6,900 (6,200–7,900)	Jarvinen and Tanner (1982)
Geometric mean								7,804	
Pomacea paludosa	FT	Meas	87.0	96	26–27.4	Mortality	1 day, 7 days	Test 1: 2,950 Test 2: 3,270 Test 3: 3,390	Call (1993)

(continued)

Table 5 (continued)

Species	Test type	Meas/ Nom	Chemical grade (%)	Duration (h)	Temp (°C)	End point	Age/size	LC/EC$_{50}$ (μg/L)	Reference
Geometric mean								3,198	
Procloeon sp.	S/R	Meas	99.0	48	22.1	Mortality	0.5–1 cm	Test 1: 1.53 Test 2: 2.11 Test 3: 1.77	Anderson et al. (2006)
Geometric mean								1.79	
Salvelinus fontinalis	FT	Meas	92.5	96	12	Mortality	1 year	Test 1: 800 (440–1,140) Test 2: 450 (320–630) Test 3: 1,050 (720–1,520)	Allison and Hermanutz (1977)
Geometric mean								723	

All studies were rated relevant and reliable (RR) and were conducted at standard temperature for a given species
S static, *SR* static renewal, *FT* flow through

Table 6 Final chronic toxicity data set for diazinon

Species	Test type	Meas/ Nom	Chemical grade (%)	Duration (days)	Temp (°C)	Endpoint	Age/size	NOEC (μg/L)	LOEC (μg/L)	MATC (μg/L)	Reference
Daphnia magna	FT	Meas	87.7	21	20	Mortality/ immobility	<24 h	0.17	0.32	0.23	Surprenant (1988)
Pimephales promelas	FT	Meas	92.5	274	25	Mortality	5 days	28	60.3	41	Allison and Hermanutz (1977)
P. promelas	FT	Meas	87.1	32	23.5–26.0	Weight	Newly hatched	50	90	67	Jarvinen and Tanner (1982)
Geometric mean										54	
Salvelinus fontinalis	FT	Meas	92.5	173	±1°C; variable acc. to date	Mortality	1 year	4.8	9.6	6.8	Allison and Hermanutz (1977)
Selenastrum capricornutum	S	Meas	87.7	7	24	Mean standing crop (cells/ mL)	6–8-day-old culture	–	–	EC_{50} 6,400	Hughes (1988)
S. capricornutum	S	Meas	87.7	7	24	Mean standing crop (cells/ mL)	6–8-day-old culture	–	–	EC_{25} 4,250	Hughes (1988)

All studies were rated relevant and reliable (RR)
S static, *S/R* static renewal, *FT* flow through

Table 7 Final acute toxicity data set for malathion

Species	Test type	Meas/ Nom	Chemical grade (%)	Duration (h)	Temp (°C)	End point	Age/size	LC_{50}/EC_{50} (μg/L)	Reference
Acroneuria pacifica	FT	Nom	95	96	12.8	Mortality	Naiads	7.7	Jensen and Gaufin (1964b)
Anisops sardeus	S	Nom	>99	48	27	Immobility/ mortality	Adult	42.2 (40.5–44.9)	Lahr et al. (2001)
Ceriodaphnia dubia	S	Nom	99.2	48	25	Mortality	≤24 h	3.35 (2.68–3.93)	Maul et al. (2006)
C. dubia	S	Nom	97	48	25	Mortality	≤24 h	1.14 (1.04–0.25)	Nelson and Roline (1998)
Geometric mean								1.95	
Chironomus tentans	S	Meas	98	96	20	Immobility/ mortality	Fourth instar	1.5 (1.2–1.9)	Belden and Lydy (2000)
C. tentans	S	Nom	99	96	20	Immobility/ mortality	Fourth instar	19.09 (11.98–30.44)	Pape-Lindstrom and Lydy (1997)
Geometric mean								5.35	
Daphnia magna	S	Nom	Analytical	48	21	Immobility/ mortality	<24 h	1.8 (1.5–2.0)	Kikuchi et al. (2000)
Elliptio icterina	S	Nom	96	96	25	Mortality	Juvenile	32,000	Keller and Ruessler (1997)
Gambusia affinis	S	Nom	>90	48	27	Mortality	5 days	3,440 (2,720–4,370)	Tietze et al. (1991)
Gila elegans	SR	Meas	93	96	22	Mortality	6 days	15,300	Beyers et al. (1994)
Jordanella floridae	FT	Meas	95	96	24.4–25.2	Mortality	33 days	349	Hermanutz (1978)
Lampsilis siliquoidea	S	Nom	96	48	25 (pH 7.5)	Mortality	Glochidia	7,000	Keller and Ruessler (1997)
Lampsilis subangulata	S	Nom	96	96	25 (pH 7.5)	Mortality	Juvenile	28,000	Keller and Ruessler (1997)
Megalonaias nervosa	S	Nom	96	24	25 (pH 7.5)	Mortality	Glochidia	22,000	Keller and Ruessler (1997)

Morone saxatilis	FT	Meas	94.2	96	15–17	Mortality	11 days	16 (13–19)	Fujimura et al. (1991)
M. saxatilis	FT	Meas	94.2	96	15–17	Mortality	45 days	25 (19–34)	Fujimura et al. (1991)
M. saxatilis	FT	Meas	94.2	96	15–17	Mortality	29 days	12 (11–14)	Fujimura et al. (1991)
M. saxatilis	FT	Meas	94.2	96	15–17	Mortality	13 days	64 (55–77)	Fujimura et al. (1991)
M. saxatilis	FT	Meas	94.2	96	15–17	Mortality	45 days	100 (87–150)	Fujimura et al. (1991)
M. saxatilis	FT	Meas	94.2	96	15–17	Mortality	45 days	66 (58–74)	Fujimura et al. (1991)
Geomean								36	
Neomysis mercedis	FT	Meas	94.2	96	17	Mortality	Neonates: ≤5 days	2.2 (2.0–2.5)	Brandt et al. (1993)
N. mercedis	FT	Meas	94.2	96	17	Mortality	Neonates: ≤5 days	1.5 (1.2–1.8)	Brandt et al. (1993)
N. mercedis	FT	Meas	94.2	96	17	Mortality	Neonates: ≤5 days	1.4 (1.3–1.5)	Brandt et al. (1993)
Geomean								1.7	
Oncorhynchus clarki	SR	Nom	95	96	13	Mortality	0.33	Test 1: 150 (133–170)	Post and Schroeder (1971)
O. clarki	SR	Nom	95	96	13	Mortality	1.25 g	Test 2: 201 (175–231)	Post and Schroeder (1971)
Geometric mean								174	
Oncorhynchus kisutch	SR	Nom	95	96	13	Mortality	1.7 g	130 (208–388)	Post and Schroeder (1971)
Oncorhynchus mykiss	SR	Nom	95	96	13	Mortality	0.41 g	122 (98–153)	Post and Schroeder (1971)
Pimephales promelas	FT	Meas	95	96	25	Mortality	29–30 days; 0.069 g; 1.7 cm	141,00 (12,300–16,100)	Geiger et al. (1984)
Pteronarcys californica	S	Nom	95	96	11.5	Mortality	Naiads, 4–6 cm	50	Jensen and Gaufin (1964a)
Ptychocheilus lucius	SR	Meas	93	96	22	Mortality	26 days	9,140	Beyers et al. (1994)

(continued)

Table 7 (continued)

Species	Test type	Meas/Nom	Chemical grade (%)	Duration (h)	Temp (°C)	End point	Age/size	LC_{50}/EC_{50} (μg/L)	Reference
Rana palustris	S	Meas	98	48	16.5	Mortality	Tadpole, Gosner 26	17,100	Budischak et al. (2009)
Salvelinus fontinalis	SR	Nom	95	96	13	Mortality	Test 1: 1.15 g	Test 1: 130 (110–154)	Post and Schroeder (1971)
S. fontinalis	SR	Nom	95	96	13	Mortality	Test 2: 2.13 g	Test 2: 120 (96–153)	Post and Schroeder (1971)
Geometric mean								125	
Simulium vittatum	S	Meas	98	48	21	Mortality	Sixth and seventh instar	54.20 (44.70–66.43)	Overmyer et al. (2003)
Streptocephalus sudanicus	S	Nom	>99	48	27	Immobility/mortality	Adult	67,750 (52,220–90,300)	Lahr et al. (2001)
Utterbackia imbecillis	S	Nom	96	96	25 (pH 7.5)	Mortality	Juvenile	215,000	Keller and Ruessler (1997)
Villosa lienosa	S	Nom	96	24	25 (pH 7.9)	Mortality	Glochidia	54,000	Keller and Ruessler (1997)
Villosa villosa	S	Nom	96	96	25 (pH 7.9)	Mortality	Juvenile	142,000	Keller and Ruessler (1997)

All studies were rated RR and were conducted at standard temperature
S static, *SR* static renewal, *FT* flow through

Table 8 Final chronic toxicity data set for malathion

Species	Test type	Meas/ Nom	Chemical grade (%)	Duration	Temp (°C)	Endpoint	Age/size	NOEC (μg/L)	LOEC (μg/L)	MATC (μg/L)	Reference
Clarias gariepinus	SR	Nom	98	5 days	27	Length/ weight	Eggs	630	1,250	887	Nguyen and Janssen (2002)
C. gariepinus	SR	Nom	98	5 days	27	Length	Eggs 3–5 hold	1,250	2,500	1,768	Lien et al. (1997)
Geometric mean										1,252	
Daphnia magna	FT	Meas	94	21 days	20	Mortality	First instar <24 h	0.06	0.1	0.077	Blakemore and Burgess (1990)
Gila elegans	FT	Meas	93	32 days	22	Growth	48 days	990	2,000	1,407	Beyers et al. (1994)
Jordanella floridae	FT	Meas	95	30 days	25.1–25.4	Growth	1–2 days	8.6	10.9	9.68	Hermanutz (1978)
Lepomis macrochirus	FT	Meas	95	10 months	9–29	Mortality	8 cm, 12 g, 1.5 years	7.4	14.6	10.4	Eaton (1970)
Oncorhynchus mykiss	FT	Meas	94	97 days	7.8–13.6	Mortality	Eggs 8 h post fert.	21	44	30.4	Cohle (1989)
Ptychocheilus lucius	FT	Meas	93	32 days	22	Growth	41 days	1,680	3,510	2,428	Beyers et al. (1994)
P. lucius	FT	Meas	93	32 days	22	Mortality	41 days	1,680	3,510	2,428	Beyers et al. (1994)
Geometric mean										2,428	

All studies were rated RR and were conducted at standard temperature
SR static renewal, *FT* flow through

and 3 SMCVs (Tables 3 and 4), the final diazinon data sets contain 13 SMAVs and 5 SMCVs (Tables 5 and 6), and the final malathion data sets contain 27 SMAVs and 7 SMCVs (Tables 7 and 8).

4 Acute Criterion Calculations

The final acute data sets for both chlorpyrifos and diazinon (Tables 3 and 5) include species from each of the five taxa requirements of the SSD procedure: a warm water fish, a species in the family Salmonidae, a planktonic crustacean, a benthic crustacean, and an insect (TenBrook et al. 2010). Cumulative probability plots of the SMAVs (Figs. 1 and 2) revealed bimodal distributions for both compounds, with invertebrates encompassing the lower subset and fish and amphibians in the upper subset. However, the SSDs were fit to the entire data set for both compounds because it is preferable to use all of the data, unless the goodness of fit test indicates a lack of fit to the entire data set. The Burr Type III SSD was fit to these data sets for the acute criteria calculations because more than eight acceptable acute toxicity values were available in the chlorpyrifos and diazinon acute data sets. The Burr Type III SSD consists of a family of three related distributions, among which the

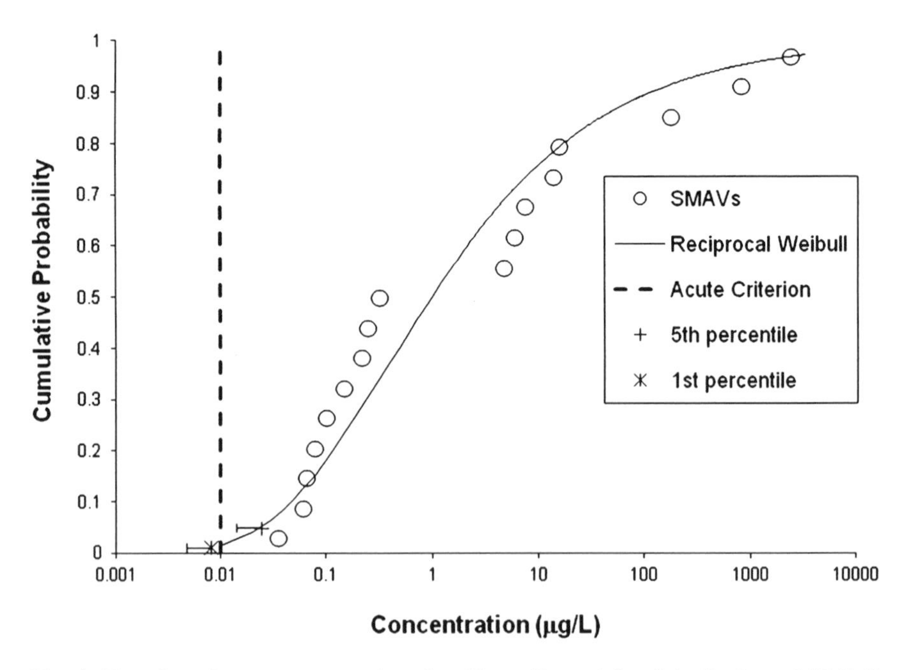

Fig. 1 Plot of species mean acute values for chlorpyrifos and fit of the Reciprocal Weibull distribution. The graph shows the median fifth and first percentiles with the lower 95% confidence limits and the acute criterion at 0.01 µg/L

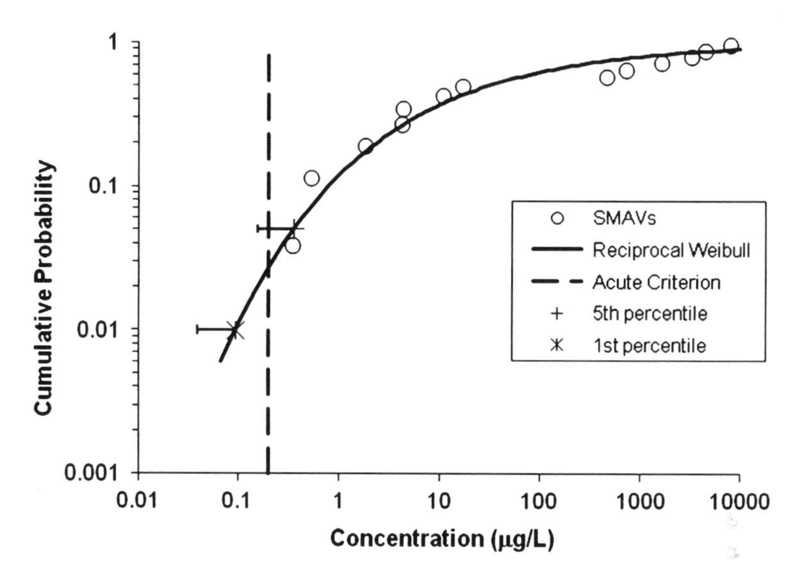

Fig. 2 Plot of species mean acute values for diazinon and fit of the Reciprocal Weibull distribution. The graph shows the median fifth and first percentiles with the lower 95% confidence limits and the acute criterion at 0.2 μg/L

BurrliOZ software (CSIRO 2001) selected the Reciprocal Weibull distribution as the best fit for both compounds based on maximum likelihood estimation.

The BurrliOZ software was used to derive fifth percentiles (median and lower 95% confidence limit), as well as first percentiles (median and lower 95% confidence limit). The median fifth percentile was used in criteria derivation because it is the most robust of the distributional estimates.

Chlorpyrifos Reciprocal Weibull Distribution

Fit parameters: $\alpha = 0.691$; $\beta = 0.394$ (likelihood = 54.083508)
Fifth percentile, 50% confidence limit: 0.0243 μg/L
Fifth percentile, 95% confidence limit: 0.0144 μg/L
First percentile, 50% confidence limit: 0.00816 μg/L
First percentile, 95% confidence limit: 0.00469 μg/L
Recommended acute value = 0.0243 μg/L (median fifth percentile)

$$\text{Acute criterion} = \frac{\text{Acute value}}{2}. \tag{1}$$

Chlorpyrifos acute criterion = 0.01 μg/L

Diazinon Reciprocal Weibull Distribution

Fit parameters: $\alpha = 2.123041$; $\beta = 0.326993$ (likelihood = 87.377508)
Fifth percentile, 50% confidence limit: 0.349 μg/L

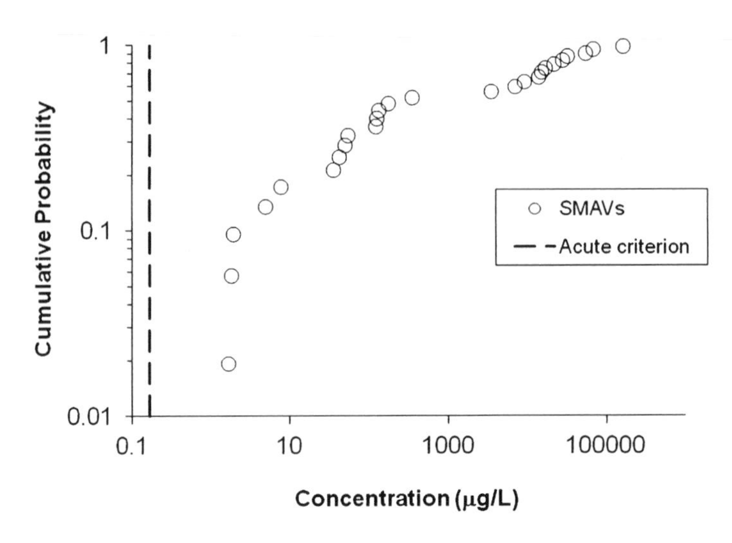

Fig. 3 Malathion species mean acute values with the acute criterion displayed at 0.17 μg/L

Fifth percentile, 95% confidence limit: 0.155 μg/L
First percentile, 50% confidence limit: 0.0937 μg/L
First percentile, 95% confidence limit: 0.0392 μg/L
Recommended acute value = 0.349 μg/L (median fifth percentile)
Diazinon acute criterion = 0.2 μg/L

No significant lack of fit to the whole data sets was found for either compound using a fit test based on cross validation and Fisher's combined test, with $X^2_{2n} = 0.1326$ for chlorpyrifos and $X^2_{2n} = 0.1561$ for diazinon (calculations shown in the Supporting Material http://extras.springer.com/). The acute data sets and corresponding Reciprocal Weibull distributions are shown in Figs. 1 and 2. The criteria are reported with one significant figure because of the variability indicated by the different confidence limit estimates.

The cumulative probability plot of the malathion SMAVs (Fig. 3) indicated that the data set is possibly bimodal, but the trend is not clearly defined. The malathion acute data set did not contain a species that fulfilled the benthic crustacean taxa requirement for use of an SSD; therefore, the malathion acute criterion could not be calculated with an SSD, and was instead calculated with an assessment factor (AF) procedure. The AF procedure estimates the median fifth percentile of the distribution by dividing the lowest SMAV in the data set by an AF, the magnitude of which was determined by the number of taxa available that fulfill the five SSD taxa requirements. An AF of 5.1 was used because the malathion data set contained four of the five taxa requirements (TenBrook et al. 2010) and the lowest SMAV in the malathion data set is 1.7 μg/L for *Neomysis mercedis*.

$$\begin{aligned}\text{Acute value} &= \frac{\text{Lowest SMAV}}{\text{Assessment factor}}, \\ &= \frac{1.7\mu\text{g/L}}{5.1} = 0.333 \quad \mu\text{g/L}.\end{aligned} \tag{2}$$

Using Eq. 1:

$$\text{Malathion acute criterion} = \frac{0.333\mu\text{g/L}}{2} = 0.17 \quad \mu\text{g/L}.$$

5 Chronic Criterion Calculations

Chronic data were limited for each of the three selected organophosphates and none of the chronic data sets contained enough data to meet the five taxa requirements of the SSD procedure. Thus, ACRs were used to calculate the chronic criteria (TenBrook et al. 2010). The UCDM ACR procedure follows the USEPA (1985) ACR instructions, except that the UCDM includes a default ACR that can be used when ACRs based on experimental data are lacking. For chlorpyrifos, two of the five SSD taxa requirements were satisfied: warm water fish (*Pimephales promelas*) and planktonic crustacean (*Ceriodaphnia dubia* and *N. mercedis*). To avoid excessive layers of estimation, the estimated chronic values for *N. mercedis* were not used to calculate ACRs, but the other two chronic data were used with appropriate corresponding acute data to calculate species mean ACRs (SMACRs). Since there were insufficient freshwater data to satisfy the three family requirements of the ACR procedure (viz., a fish, an invertebrate, and another sensitive species), saltwater data for the California grunion (*Leuresthes tenuis*) were used to meet the third taxa requirement. Three of the five diazinon taxa requirements were satisfied: a species in the family Salmonidae (*Salvelinus fontinalis*), a warm water fish (*P. promelas*), and a planktonic crustacean (*Daphnia magna*). These three chronic values were each paired with appropriate corresponding acute toxicity values, which satisfied the three family requirements for the ACR procedure. Three of the malathion chronic toxicity values were paired with corresponding acute toxicity values (*Gila elegans*, *Ptychocheilus lucius*, *Jordanella floridae*). Since only fish data were available, the invertebrate taxa requirement was not satisfied. A default ACR of 12.4 was included in the malathion ACR data set to compensate for the lack of invertebrate data (TenBrook et al. 2010).

An SMACR was calculated by dividing the acute LC_{50} by the chronic maximum acceptable toxicant concentration (MATC) for a given species (Tables 9–11). The final ACR for malathion of 11.8 was calculated as the geometric mean of all the SMACRs in the data set and one default ACR (Table 11). The SMACRs varied by

Table 9 Calculation of the final acute-to-chronic ratio for chlorpyrifos

Species	LC_{50} (μg/L)	Reference	Chronic end point	MATC (μg/L)	Reference	ACR (LC_{50}/MATC)
Ceriodaphnia dubia	0.0396	CDFG (1999)	Mortality	0.040	CDFG (1999)	1.0
C. dubia	0.0396	CDFG (1999)	Reproduction	0.040	CDFG (1999)	1.0
C. dubia				Species mean ACR		1.0[a]
Pimephales promelas	140	Jarvinen and Tanner (1982)	Weight	2.3	Jarvinen and Tanner (1982)	61[b]
Leuresthes tenuis[c]	1.0	Borthwick et al. (1985)	Growth	0.2	Goodman et al. (1985)	5.0[a]
					Final ACR	2.2

[a] Values used in calculation
[b] Excluded; >10× the ACR for cladocerans whose species mean acute value is nearest the fifth percentile of 0.026 μg/L
[c] Saltwater species included in ACR calculation; study rated relevant and reliable in every other respect

Table 10 Calculation of the species mean acute-to-chronic ratios for diazinon

Species	LC_{50} (μg/L)	Chronic end point	MATC (μg/L)	Reference	ACR (LC_{50}/MATC)
Daphnia magna	0.52	21 days mortality/immobility	0.23	Surprenant (1988)	2.3[a]
Pimephales promelas	7,800	274 days mortality	41	Allison and Hermanutz (1977)	190
P. promelas	6,900	32 days weight	67	Jarvinen and Tanner (1982)	103
P. promelas				Species mean ACR	140[b]
Salvelinus fontinalis	723	173 days mortality	6.8	Allison and Hermanutz (1977)	106[b]

[a] Value used in calculation
[b] Excluded; >10× the ACR for cladocerans whose species mean acute value is nearest the fifth percentile of 0.026 μg/L

more than a factor of 10, and there was an increasing trend of SMACRs as the SMAVs increased for both chlorpyrifos and diazinon. To utilize the most relevant values for these two compounds, the final multispecies ACRs were calculated as the geometric mean of the SMACRs for species whose SMAVs were close to the acute value. For chlorpyrifos, the species with an SMAV closest to the acute median fifth percentile was *C. dubia* (SMAV = 0.0396 μg/L), with an SMACR

Table 11 Calculation of the final acute-to-chronic ratio for malathion

Species	LC_{50} (μg/L)	Reference	Chronic end point	MATC (μg/L)	Reference	ACR (LC_{50}/MATC)
Gila elegans	15,300	Beyers et al. (1994)	Growth	1,407	Beyers et al. (1994)	10.8
Jordanella floridae	349	Hermanutz (1978)	Growth	9.68	Hermanutz (1978)	36.0
Ptychocheilus lucius	9,140	Beyers et al. (1994)	Growth	2,428	Beyers et al. (1994)	3.7
Invertebrate	Default ACR					12.4
					Final ACR	11.8

All values were used in the calculation

of 1.0. The SMACR for *L. tenuis* was within a factor of 10 of this, so it was also included in the calculation, to give a final ACR of 2.2 for chlorpyrifos. The species with an SMAV closest to the acute median fifth percentile for diazinon was *D. magna* (SMAV = 0.52 μg/L), with an SMACR of 2.3. None of the other SMACRs were within a factor of 10 of this value; therefore, the final multispecies ACR was 2.3 for diazinon. To calculate the chronic criteria, the recommended acute values (median fifth percentiles) were divided by the final ACRs. The diazinon chronic criterion is adjusted downward later in this chapter based on comparisons to data for sensitive species, threatened and endangered species, and ecosystem-level effects.

Chlorpyrifos chronic criterion calculated with the acute median fifth percentile estimate:

Fifth percentile, 50% confidence limit: 0.0243 μg/L

$$\begin{aligned}\text{Chronic criterion} &= \frac{\text{Acute fifth percentile}}{\text{ACR}},\\ &= \frac{0.0243\ \mu\text{g/L}}{2.2},\\ &= 0.01\ \ \mu\text{g/L}.\end{aligned}$$

Diazinon chronic criterion calculated with the acute median fifth percentile estimate:

Fifth percentile, 50% confidence limit: 0.349 μg/L

$$\begin{aligned}\text{Chronic criterion} &= \frac{0.349\mu\text{g/L}}{2.3},\\ &= 0.2\ \ \mu\text{g/L}.\end{aligned}$$

Malathion chronic criterion calculated with the acute median fifth percentile estimate:

Fifth percentile, 50% confidence limit: 0.333 μg/L

$$\begin{aligned}\text{Chronic criterion} &= \frac{0.333\,\mu\text{g/L}}{11.80},\\ &= 0.028\ \ \mu\text{g/L}.\end{aligned}$$

6 Bioavailability

Chlorpyrifos, diazinon, and malathion have moderate to high octanol–water partition coefficients (log K_{ow}s of 4.96, 3.81, and 2.84, respectively), indicating that sorption to sediment or dissolved organic matter could reduce bioavailability of these compounds, but few studies were identified regarding this topic. One relevant study reported that the bioavailability of diazinon to *D. magna* was inversely proportional to the dissolved humic material concentration, presumably because diazinon was binding to the dissolved humic material (Steinberg et al. 1993). The results of a study by Phillips et al. (2003) are less clear; they found that fewer walleye survived exposure to chlorpyrifos–humic acid (HA) complexes than to either HA alone or chlorpyrifos alone, and no differences were seen in cholinesterase inhibition between chlorpyrifos–HA and aqueous chlorpyrifos exposures. The uptake of malathion from spiked sediment by freshwater snails (*Stagnicola* sp.) occurred quickly (up to 0.1 μg/g in 36 h), indicating that malathion was bioavailable in sediment (Martinez-Tabche et al. 2002). With such little and inconsistent information regarding the toxicity of the three selected organophosphates when bound or complexed, the bioavailability of these compounds is not predictable without site-specific, species-specific data. Until such data is available, it is recommended that criteria compliance should be determined based on whole water concentrations.

7 Chemical Mixtures

Mixtures of OP pesticides are common in waterways of the USA (Gilliom 2007) and several studies have demonstrated that mixtures of organophosphates exhibit additive toxicity (Bailey et al. 1997; Hunt et al. 2003; Lydy and Austin 2005; Rider and LeBlanc 2005). Because all OPs have the same mode of action, concentration addition is a valid assumption. To determine criteria compliance when a mixture of OPs is present, either the toxic unit or relative potency factor approach can be used (TenBrook et al. 2010). However, concentration addition may underestimate mixture toxicity of OPs in some cases. For example, malathion had a synergistic, rather than additive effect on acetylcholinesterase (AChE) activities in Coho salmon (*Oncorhynchus kisutch*; Laetz et al. 2009) when combined with either chlorpyrifos or diazinon. Many fish species die after high rates of acute brain AChE inhibition (>70–90%; Fulton and Key 2001), but this study did not provide a way to quantitatively incorporate these nonadditive interactions into compliance.

Several researchers reported greater than additive toxicity of both chlorpyrifos and diazinon in combination with triazine herbicides (Anderson and Lydy 2002; Belden and Lydy 2000; Jin-Clark et al. 2002; Lydy and Austin 2005) while additive effects were reported for a mixture of atrazine and malathion (Belden and Lydy 2000). Multiple interaction coefficients (also called synergistic ratios) were available for atrazine with either chlorpyrifos or diazinon over a range of concentrations, so these values were used to derive quantitative relationships. The interaction coefficient (K) is calculated by dividing the concentration that affects 50% of the exposed population (EC_{50}) for the pesticide alone by the EC_{50} in the presence of a nontoxic concentration of the synergist. When K is greater than unity, a synergistic interaction is indicated, and when K is less than unity an antagonistic interaction is indicated. All available Ks for chlorpyrifos and diazinon are given in Tables S4 and S5 (Supporting Material http://extras.springer.com/).

Least squares regressions of the *Chironomus tentans* and *Hyalella azteca* combined diazinon data resulted in a significant relationship between atrazine concentration and K ($p<0.001$; JMP IN v.5.1.2; JMP 2004):

$$K = 0.0095\ [\text{atrazine}] + 1.05\ (r^2 = 0.87, \text{p} = 0.0007).$$

To determine compliance or to assess potential for harm, Eq. 4 may be used to establish the effective concentration of diazinon in the presence of atrazine:

$$C_\text{a} = C_\text{m}\ (K), \tag{3}$$

where C_a is the adjusted, or effective, concentration of chemical of concern (i.e., diazinon); C_m is the concentration measured for chemical of concern (i.e., diazinon); and K is the coefficient of interaction, calculated for the synergist concentration in water.

The effective concentration may be compared to diazinon criteria or may be used in one of the additivity models.

Least squares regressions of the combined *C. tentans* and *H. azteca* chlorpyrifos data also resulted in a significant relationship between atrazine concentration and Ks ($p<0.005$; JMP IN v.5.1.2; JMP 2004), but the r^2 is not very high ($r^2 = 0.52$); so the two species were considered independently. For *C. tentans*, the relationship between K and atrazine concentration was not significant ($p>0.05$), but for *H. azteca* the following relationship was determined:

$$K = 0.009\ [\text{atrazine}] + 1.12\ (r^2 = 0.94, \text{p} = 0.03).$$

This relationship should be used with caution because of the small data set ($n = 4$) and the fact that three of the four values are from the same study. The lack of a significant relationship between atrazine concentration and *K*s for *C. tentans* may be due to differences between studies (there were not enough data to evaluate the experiment effect statistically). Since *H. azteca* is among the most sensitive species in the data set, it is worthwhile to use Eq. 4 to estimate *K*s for various levels of atrazine co-occurring with chlorpyrifos. To assess potential for harm, Eq. 4 may be used to estimate the effective concentration of chlorpyrifos in the presence of atrazine, which may be compared to chlorpyrifos criteria or may be used in one of the additivity models.

The toxicity of mixtures of chlorpyrifos, diazinon, and/or malathion has been documented to occur with many other chemicals (Ankley and Collyard 1995; Bailey et al. 2001; Banks et al. 2003, Belden and Lydy 2006; Denton et al. 2003; Hermanutz et al. 1985; Macek 1975; Mahar and Watzin 2005; Overmyer et al. 2003; Rawash et al. 1975; Solomon and Weis 1979; Van Der Geest et al. 2000; Venturino et al. 1992), but multispecies synergistic ratios are not available; so these interactions cannot be incorporated into criteria compliance.

8 Water Quality Effects

Several studies have shown increased toxicity of chlorpyrifos and diazinon with increased temperature (Humphrey and Klumpp 2003; Johnson and Finley 1980; Landrum et al. 1999; Lydy et al. 1999; Macek et al. 1969; Mayer and Ellersieck 1986; Patra et al. 2007). Conversely, one toxicity study on malathion demonstrated decreased toxicity with increasing temperature due to increased degradation of malathion (Keller and Ruessler 1997). However, none of these studies were rated RR, so they were not used to quantify effects of temperature on toxicity in criteria compliance. In addition, two studies showed no effect of pH on toxicity (Keller and Ruessler 1997; Landrum et al. 1999).

9 Sensitive Species

The criteria derived using the acute median fifth percentiles were compared to toxicity values for the most sensitive species in both the acceptable (RR) and supplemental (RL, LR, LL) data sets (Tables S6–S8, Supporting Material http://extras.springer.com/) to ensure that all species are adequately protected in an ecosystem. The malathion criteria are below all available toxicity data, so there is no indication of underprotection of sensitive species in the data set. There is one measured chlorpyrifos chronic value that is just under the derived chronic criterion, which is an MATC of 0.0068 μg/L for *Mysidopsis bahia* (Sved et al. 1993); however, this is a saltwater species and there were significant effects observed in the solvent control.

The estimated chronic value of 1 ng/L for *N. mercedis* (CDFG 1992a, d) is below the calculated criterion, but the chronic criterion should not be adjusted unless the estimated value is supported by measured data.

The lowest value in the acute diazinon RR data set is a value for *C. dubia* of 0.21 μg/L (Table 5), which is almost identical to the calculated criterion of 0.2 μg/L. This value for *C. dubia* is the lowest compared to ten others used for criteria derivation (0.26, 0.29, 0.32, 0.33, 0.33, 0.35, 0.38, 0.436, 0.47, 0.507, SMAV is 0.34 μg/L). There is also a similar value in the supplemental data set of 0.25 μg/L (Table S7, Supporting Material http://extras.springer.com/). In this case, downward adjustment of the acute criterion is not recommended because the *C. dubia* SMAV of 0.34 μg/L indicates that the acute criterion of 0.2 μg/L is protective of this species.

The lowest measured SMCV in the diazinon data set rated RR is 0.23 μg/L for *D. magna* (Surprenant 1988), which is just above the chronic criterion (0.2 μg/L). This is the only highly rated value for *D. magna* or any cladoceran species. The supplemental data set (Table S7, Supporting Material http://extras.springer.com/) contains 6 MATCs for *D. magna* that are approximately equivalent to the criterion (0.16, 0.16, 0.22, 0.24, 0.24, and 0.24 μg/L; Dortland 1980; Fernández-Casalderrey et al. 1995; Sánchez et al. 1998) and 12 MATCs for *D. magna* of 0.07 μg/L that are below the chronic criterion (Sánchez et al. 1998, 2000). These studies did not rate highly because test parameters were not well-documented, but had no obvious flaws in study design or execution. Sánchez et al. (2000) reported the concentrations incorrectly in their original report as ng/L instead of μg/L, which was confirmed via correspondence with the authors. This was a multigenerational test, which would be expected to be more sensitive than the test rated RR that only monitored reproduction in one generation (Surprenant 1988). The only other chronic value for a cladoceran is 0.34 μg/L for a *C. dubia* 7-day test (Norberg-King 1987) in the supplemental data set. *C. dubia* is the most sensitive species in the acute distribution; thus, this gap in the RR chronic data set may lead to an underprotective criterion. The supplemental data set also contains a toxicity value of 0.13 μg/L for *H. azteca*, which is below the chronic criterion, but the end point in this study does not have an established connection to survival, growth, or reproduction.

Based on this evidence, the diazinon chronic criterion, as calculated, may be underprotective of cladocerans; therefore, the next lowest distributional estimate was used to calculate the chronic criterion. Using the lower 95% confidence limit of the fifth percentile to calculate the chronic criterion yielded a recommended chronic criterion of 0.07 μg/L for diazinon.

Diazinon chronic criterion calculated with the lower 95% confidence interval of the acute fifth percentile estimate:

Fifth percentile, lower 95% confidence limit: 0.155 μg/L

$$\begin{aligned}\text{Chronic criterion} &= \frac{0.155\,\mu\text{g/L}}{2.3},\\ &= 0.07\ \ \mu\text{g/L}.\end{aligned}$$

10 Ecosystem-Level Studies

Multispecies studies may provide more realistic exposure conditions than single-species laboratory studies; therefore, the results of these studies were compared to the derived chronic criteria to ensure that the criteria are protective of ecosystems. Twenty-one chlorpyrifos studies, four diazinon studies, and two malathion studies on the effects on microcosms, mesocosms, and model ecosystems were rated acceptable (R or L reliability rating, Table S9, Supporting Material http://extras.springer.com/). In the two acceptable malathion studies, the authors applied concentrations well above the chronic criterion and did not calculate ecosystem-level NOECs (Kennedy and Walsh 1970; Relyea 2005); thus, no information was reported by these authors that indicates that the chronic malathion criterion is underprotective of organisms in ecosystems.

Many of the chlorpyrifos studies involved one-time application at levels well above the calculated criteria (Brock et al. 1992a, b, 1993; Cuppen et al. 1995; Kersting and Van Wijngaarden 1992; Rawn et al. 1978; Van Breukelen and Brock 1993; Van Donk et al. 1995; Van Wijngaarden and Leeuwangh 1989). The authors of several other chlorpyrifos studies reported effects with exposures ranging from 0.1 to 2 μg/L, which are 1–2 orders of magnitude higher than the derived criteria (Eaton et al. 1985; Giddings et al. 1997; Macek et al. 1972; Pusey et al. 1994; Van Den Brink et al. 1995; Van Wijngaarden 1993; Ward et al. 1995). Four studies provided community NOECs for chlorpyrifos, which are the most relevant values to compare to the derived chronic criterion (0.01 μg/L). Van Wijngaarden et al. (1996) reported 7-day mesocosm EC_{50}s ranging from 0.1 μg/L for *Mystacides* spp. to 2.8 μg/L for *Ablabesmyia* spp. In the same study, 7-day EC_{10}s were reported, which are sometimes equated to MATCs, and the EC_{10}s values ranged from 0.01 μg/L for *Mystacides* spp. to 2.7 μg/L for *Ablabesmyia* spp. indicating that the chronic criterion would likely be protective of *Mystacides* spp. Van Wijngaarden et al. (2005) and Van Den Brink et al. (1996) both reported community NOECs of 0.1 μg/L in laboratory microcosms and outdoor experimental ditches. In various measures of ecosystem metabolism, Kersting and Van Den Brink (1997) reported ecosystem NOECs ranging from <0.1 to 6 μg/L chlorpyrifos based on system oxygen concentration, system pH, gross production (mg O_2/L-d), and respiration (mg O_2/L-d). The authors acknowledged that the latter two significant findings may be due to a Type II error.

Werner et al. (2000) performed laboratory toxicity tests and toxicity identification evaluations on samples collected from the Sacramento-San Joaquin River Delta. Six filtered samples exhibiting significant mortality in $\leq$4 days had chlorpyrifos concentrations ranging from 0.09 to 0.52 μg/L (with no other pesticides detected). Two filtered samples exhibiting chronic toxicity (significant mortality in >4 days) had chlorpyrifos concentrations ranging from 0.058 to 0.068 μg/L (with no other pesticides detected). Hundreds of other samples did not exhibit toxicity, implying that they had chlorpyrifos levels below those found in the samples that induced toxicity. In a treated pond study by Siefert (1984), the first two applications of a granular formula resulted in variable measured chlorpyrifos concentrations ranging from nondetects to 0.30 μg/L and reduction or elimination of seven species of cladocerans and benthic invertebrates. Unfortunately, there is no way to determine the no-effect concentration in this study. However, one of the most sensitive species in the study was *H. azteca*, which was included in the criteria derivation. Given the results of these studies, it appears that acute and chronic criteria of 0.01 μg/L are protective of organisms in ecosystems.

The four acceptable diazinon ecosystem studies did not indicate that the derived criteria are underprotective of any tested species. Giddings et al. (1996) applied a range of diazinon concentrations (2.0–500 μg/L) to aquatic microcosms and reported a community-level LOEC of 9.2 μg/L and a community-level NOEC of 4.3 μg/L (70-day averages). Arthur et al. (1983) used three outdoor experimental channels to assess the effect of a 12-week exposure to diazinon using a low treatment of 0.3 μg/L and high treatment of 6 μg/L (nominal concentrations), followed by 4 week at higher concentrations (12 and 30 μg/L, respectively). Effects on amphipods and insects were seen in the lowest treatment with lower numbers of mayflies and damselflies emerging from treated channels. Moore et al. (2007) reported that survival of *H. azteca* was affected after exposure to leaf litter contaminated with diazinon (measured residues of $\geq$60 μg/kg). The concentrations tested in these ecosystem studies are all well above the diazinon criteria, except the study by Arthur et al. (1983) that documented effects at 0.3 μg/L, which is only slightly above the chronic criterion derived using the acute median fifth percentile (0.2 μg/L). This study adds support for use of a lower chronic criterion of 0.07 μg/L (derived using the lower 95% confidence interval of the acute fifth percentile).

11 Threatened and Endangered Species

The derived criteria were compared to measured and predicted toxicity values for threatened and endangered species (TES), ensuring that they are protective of these species. TES were those plants and animals listed by the US Fish and Wildlife Service (USFWS 2010) and the California Department of Fish and Game (CDFG 2010a, b).

Two listed salmonid species, *Oncorhynchus mykiss* and *Oncorhynchus tshawytscha*, were included in the acute chlorpyrifos criterion calculation and their SMAVs were well above the final criterion. None of the listed animals or plants are represented in the acceptable acute or chronic diazinon data sets. There are six threatened or endangered species in the acute malathion data set: *G. elegans*, *Lampsilis subangulata*, *Oncorhynchus clarki*, *O. kisutch*, *O. mykiss*, and *P. lucius*. Three of these species are also included in the chronic malathion data set: *G. elegans*, *O. mykiss*, and *P. lucius*. The toxicity values for all of these species are at least two orders of magnitude larger than the derived malathion acute and chronic criteria, indicating that the criteria should be protective of these species.

The supplemental data sets (Tables S6–S8, Supporting Material http://extras.springer.com/) also contain toxicity values for several TES. The chlorpyrifos supplemental data set contains toxicity values for additional listed fish, *O. clarki*, *Notropis mekistocholas*, and *Gasterosteus aculeatus*, which has a listed subspecies (*G. aculeatus williamsoni*). The diazinon supplemental data set contains toxicity values for *N. mekistocholas* and two additional salmonids, *O. clarki* and *O. tshawytscha*, that are all much higher than the derived criteria. Although not as reliable, these data support that the derived criteria are protective of these endangered fish.

Toxicity data for species in the same genus or family as TES were used as surrogates to predict TES toxicity values with the USEPA interspecies correlation estimation software (Web-ICE v. 3.1; Raimondo et al. 2010). *P. promelas* was used as a surrogate to predict toxicity values for 26 TES in the Cyprinidae family and *O. mykiss* and *O. tshawytscha* were used to predict toxicity values for 11 salmonids for chlorpyrifos (Table S10, Supporting Material http://extras.springer.com/). *Gammarus pseudolimnaeus*, *S. fontinalis*, and *P. promelas* were used to predict toxicity values for a total of 41 TES for diazinon (Table S11, Supporting Material http://extras.springer.com/). For malathion, *G. elegans*, *P. promelas*, *P. lucius*, *O. clarki*, *O. kisutch*, *O. mykiss*, and *S. fontinalis* were all used as surrogates (Table S12, Supporting Material http://extras.springer.com/). Based on the available data and estimated values for animals, there is no evidence that the calculated acute and chronic criteria for chlorpyrifos, diazinon, or malathion are underprotective of TES. However, a caveat is that no data were found for effects on federally endangered cladocerans or insects, or acceptable surrogates (i.e., in the same family), which are the most sensitive species in the data sets.

There was one algal study (the only plant value) that rated RR for diazinon, but no algae species are on the federal endangered, threatened, or rare species lists. For chlorpyrifos and malathion, none of the plant studies identified rated RR, and none of the studies were for plants on the state or federal endangered, threatened, or rare species lists. Plants are relatively insensitive to OPs, so the calculated criteria should be protective of this taxon.

12 Bioaccumulation

Bioaccumulation is defined as accumulation of chemicals in an organism from all possible exposure routes, e.g., partitioning from the water and/or intake via food. A bioaccumulation factor (BAF) is a measure of the total accumulation by all possible exposure routes and is defined here as the ratio of the concentration in an organism and the concentration in surrounding media ($BAF = C_{organism}/C_{media}$). When the chemical accumulates up the food chain from prey to predator, the phenomenon is called biomagnification. The potential for bioaccumulation was assessed to ensure that if concentrations of the selected OPs are at or below the derived water quality criteria, they will not lead to toxicity in terrestrial wildlife via bioaccumulation.

Chlorpyrifos and diazinon have similar physical–chemical characteristics, including molecular weights <1,000 and log-normalized octanol–water partition coefficients (log K_{ow}) >3.0 L/kg, which indicates that both compounds have the potential to bioaccumulate. Malathion has a lower log K_{ow} of 2.84 L/kg and it does not appear to bioaccumulate from the available studies, so bioaccumulative potential was not assessed for malathion. Assessment for bioaccumulation in humans was not done because there is low potential and there are no tolerances or US Food and Drug Administration (USFDA) action levels for any of the three compounds in fish tissue (USFDA 2000).

Uptake of chlorpyrifos and diazinon from water has been measured in a number of studies and bioconcentration factors (BCFs) vary widely among different species (Table S13, Supporting Material http://extras.springer.com/). Most studies disclosed that diazinon is relatively quickly eliminated from tissues after placing organisms in clean water (3–8 days), and that a steady state is reached within a few days (Deneer et al. 1999; El Arab et al. 1990; Kanazawa 1978; Keizer et al. 1991; Palacio et al. 2002; Sancho et al. 1993; Tsuda et al. 1990, 1995, 1997). Varó et al. (2002) reported biomagnification factors (BMFs), which are a measure of uptake from food items or prey, of 0.7–0.3 (decreasing with increasing time of exposure) for chlorpyrifos in a two-level food chain experiment with *Artemia* spp., and the fish *Aphanus iberius*. BMFs of less than 1.0, and the fact that the BMFs decrease over time, indicate that chlorpyrifos does not biomagnify. Varó et al. (2002) suggest that this is due to the ability of fish to biotransform chlorpyrifos and to the moderate log K_{ow} of chlorpyrifos. Data suggests only slight bioaccumulation of malathion (Forbis and Leak 1994; Kanazawa 1975; Olvera-Hernandez et al. 2004; Tsuda et al. 1989, 1990). For the freshwater snail (*Stagnicola* sp.), uptake of malathion occurred quickly (up to 0.1 μg/g in 36 h); however, the short elimination half-life ($t_{1/2\_e} = 46.79$ h) led to the conclusion that this compound was not being stored in snails (Martinez-Tabche et al. 2002).

Since chlorpyrifos and diazinon have properties indicating bioaccumulative potential, the aqueous concentrations of these compounds required to cause toxicity due to bioaccumulation in mallard ducks (Table S14, Supporting Material http://extras.springer.com/) was estimated, and then compared to the derived criteria. For diazinon, no BAFs or BMFs were identified in the literature. A BAF can be calculated as the product of a BCF and a BMF (BAF = BCF × BMF). For diazinon, a BCF of 188 L/kg for *Poecilia reticulata* (Keizer et al. 1993) and a default BMF of 2, based on the log K_{ow} of diazinon (TenBrook et al. 2010), were used to estimate a BAF. A conservative aqueous NOEC was calculated by dividing the lowest dietary NOEC for mallard duck (8.3 mg/kg feed; USEPA 2004a) by the estimated BAF.

$$\mathrm{NOEC}_{\mathrm{water}} = \frac{\mathrm{NOEC}_{\mathrm{oral\_predator}}}{\mathrm{BCF}_{\mathrm{food\_item}} \times \mathrm{BMF}_{\mathrm{food\_item}}}. \tag{4}$$

The resulting $NOEC_{water}$ for diazinon is 22.1 μg/L, which is well above the chronic criterion of 0.07 μg/L, which indicates that diazinon at concentrations equal to or below the chronic criterion will not likely cause harm via bioaccumulation.

A similar calculation was performed with chlorpyrifos data. The highest nonlipid-based BCF (1,700 L/kg; Jarvinen et al. 1983), the highest reported BMF for chlorpyrifos of 0.7 (Varó et al. 2002), and the lowest dietary NOEC for a mallard of 25 mg/kg (USEPA 2002) were used in this analysis to assess a worst-case bioaccumulation scenario. The $NOEC_{water}$ estimated for chlorpyrifos using this data was 21 μg/L. This value is well above both the acute and chronic criteria of 0.01 μg/L; therefore, the criteria are likely to be protective of terrestrial animals feeding on aquatic organisms.

13 Harmonization with Air or Sediment Criteria

The maximum allowable concentration of these compounds in water may impact life in other environmental compartments through partitioning. Chlorpyrifos, diazinon, and malathion have all been observed in the atmosphere and shown to be transported via rain and fog (Charizopoulos and Papadopoulou-Mourkidou 1999; Glotfelty et al. 1990; McConnell et al. 1998; Scharf et al. 1992; Zabik and Seiber 1993). However, there are no federal or California state air quality standards for any of the compounds (CARB 2010; USEPA 2009b), so no estimates of the partitioning from water to the atmosphere were made. There are sediment guidelines available for diazinon and malathion that were estimated based on equilibrium partitioning from water using the USEPA water quality criteria (USEPA 2004b); these values are not useful for estimating back to a water concentration because that would simply undo the original partitioning estimate. No other federal or California state sediment quality standards were identified for these compounds (CDWR 1995; Ingersoll et al. 2000; NOAA 1999; USEPA 2009a); thus, partitioning between water and sediment was not predicted for the water quality criteria.

14 Assumptions, Limitations, and Uncertainties

The assumptions, limitations, and uncertainties involved in criteria generation are included to inform environmental managers of the accuracy and confidence in criteria. The UCDM discusses these points for each section as different procedures were chosen and includes a review of all of the assumptions inherent in the methodology (TenBrook et al. 2010). Additionally, the different calculations of distributional estimates for chlorpyrifos and diazinon included in Sect. 4 of this article may be used to consider the uncertainty in the resulting acute criteria.

For all three compounds, a major limitation was lack of chronic data, especially for the most sensitive species, cladocerans and other invertebrates. For malathion, there were inadequate invertebrate data for the ACR, so a default value was included. For diazinon, the chronic criterion calculated with the ACR and acute median fifth percentile estimate was not clearly protective of sensitive invertebrates, so the next lowest distributional estimate was used to adjust the criterion downward. Another major limitation was that the malathion acute data set was lacking the benthic crustacean taxa requirement, which precluded the use of an SSD. Instead, the final acute criterion was derived using an assessment factor. When additional highly rated data is available, particularly chronic data for invertebrates, or data regarding temperature effects or mixtures, the criteria should be recalculated to incorporate new research.

15 Comparison to Existing Criteria

There are existing state and federal water quality criteria or objectives for both chlorpyrifos and diazinon to which the criteria derived in this article can be compared. The USEPA and the CDFG have both derived water quality criteria for chlorpyrifos and diazinon using the USEPA (1985) method. The agencies derived criteria at different times, and therefore used different data sets; so the results are not identical. The USEPA (1985) criteria derivation method has been the standard used in the USA, and produces robust and reliable criteria, partly because of the large amount of data required to derive criteria following this method. One goal of creating the UCDM was to create a methodology for use in the future that had less data requirements and more flexible statistical methods than those used by the USEPA method, but which still produced criteria that are as robust and reliable as those produced by the USEPA (1985) methodology.

The final UCDM acute and chronic chlorpyrifos criteria (both 0.01 μg/L) are lower than those derived by the USEPA (1986a) of 0.084 and 0.041 μg/L, respectively, but are closer to those derived by the CDFG of 0.025 and 0.015 μg/L, respectively (Siepmann and Finlayson 2000). These three acute and chronic criteria all differ by less than a factor of 10, but there are four SMAVs in the UCDM acute data set that are below the USEPA acute criterion, and one SMCV below the USEPA chronic criterion, indicating that these species would not be protected by the USEPA criteria. After a detailed comparison of the data sets and calculation methodologies used by the different agencies (Appendix A, Supporting Material http://extras.springer.com/), it was concluded that the primary cause of differing results was the inclusion of studies performed at later dates, as described above.

The final UCDM diazinon acute criterion of 0.2 μg/L is slightly higher than the USEPA diazinon acute criterion of 0.17 μg/L (USEPA 2005) while the final UCDM diazinon chronic criterion of 0.07 μg/L is lower than the USEPA chronic criterion of 0.17 μg/L (USEPA 2005). The CDFG acute and chronic water quality criteria (0.16 and 0.10 μg/L, respectively) are also very similar to those calculated using the UCDM (Siepmann and Finlayson 2000). The acute criteria from the USEPA, the CDFG, and the UCDM all differ by less than a factor of 2, and part of the difference is because only one significant figure was reported by the UCDM while two are reported by the USEPA and the CDFG. Based on the UCDM data sets, the diazinon criteria from the various agencies all appear to be protective of aquatic ecosystems. Criteria calculated using the UCDM and the EPA method are likely similar because the criteria calculation procedures for chemicals that have larger data sets are similar in the two methods. Many of the novel aspects to the UCDM were added to enable criteria generation for compounds with more limited data sets or to incorporate other factors that affect toxicity.

In the USA, the only existing aquatic life water quality criterion identified for malathion was not derived using the USEPA (1985) methodology. Instead, a chronic criterion of 100 ng/L was calculated for malathion by applying an application factor of 0.1 to the 96-h LC_{50} data for the most sensitive species (*Gammarus lacustris*, *Gammarus fasciatus*, and *Daphnia pulex*), which were approximated as 1,000 ng/L (USEPA 1986b). This EPA chronic criterion is approximately a factor of 3.6 greater than the UCDM chronic criterion of 28 ng/L. The EPA chronic criterion would not be protective of the most sensitive species in the current UCDM data set, *D. magna* (MATC = 77 ng/L).

The UCDM criteria were also compared to criteria, or analogous values, derived by other countries. Maximum permissible concentrations (MPCs) of 0.0028, 0.037, and 0.013 μg/L for chlorpyrifos, diazinon, and malathion, respectively, were derived in the Netherlands using a statistical extrapolation method (Crommentuijn et al. 2000). MPCs are analogous to chronic criteria, and these MPCs are all lower than the UCDM chronic criteria for these compounds, which may, in part, be because the Dutch method uses NOECs instead of MATCs in their distribution. There are short-term (acute) and long-term (chronic) Canadian water quality guidelines for the protection of aquatic life for chlorpyrifos of 0.02 and 0.002 μg/L, respectively (CCME 2008). The short-term guideline was derived using an SSD while the long-term guideline was derived by applying a safety factor of 20 to the lowest acute toxicity value (0.04 μg/L for *H. azteca*). This safety factor may be overprotective because paired acute and chronic data indicate that acute and chronic toxicity occur at similar concentrations. The UK has existing environmental quality standards for diazinon, and also newly proposed values (UKTAG 2008). The existing short-term (acute) and long-term (chronic) environmental quality standards are 0.1 and 0.03 μg/L, respectively, while the proposed values are 0.02 and 0.01 μg/L, respectively. The proposed short-term value was derived by applying a safety factor of 10 to the lowest LC_{50} of 0.2 μg/L for *G. fasciatus* and the proposed long-term value was derived by applying an assessment factor of 10 to the NOEC of 0.1 μg/L for Atlantic salmon. Both the existing and proposed environmental quality standards are lower than those derived via the UCDM, but it appears that they used data not included in the UCDM data sets.

16 Comparison to the USEPA 1985 Method

The main cause for differences between criteria derived by different agencies is that different data sets were used, primarily because more studies are undertaken and completed as time passes. To compare only the SSD calculation methods, example criteria were generated for chlorpyrifos, diazinon, and malathion using the USEPA (1985) criteria derivation methodology with the data set gathered for this article. The USEPA acute methods have three additional taxa requirements beyond the five required by the SSD procedure of the UCDM. They are:

1. A third family in the phylum Chordata (e.g., fish, amphibian)
2. A family in a phylum other than Arthropoda or Chordata (e.g., Rotifera, Annelida, Mollusca)
3. A family in any order of insect or any phylum not already represented

These three additional requirements were all met for diazinon and example criteria are calculated below. The chlorpyrifos data set does not contain a family in a phylum other than Arthropoda or Chordata. However, the CDFG has calculated criteria for compounds with incomplete data sets if the missing taxa requirements are known to be relatively insensitive to the compound of interest. Data in the supplemental data set shows that mollusks are relatively insensitive to chlorpyrifos exposure (LC_{50}s>94 μg/L), so example criteria were calculated. The three additional taxa requirements were met for malathion, but the malathion data set does not contain a benthic invertebrate; so it is still deficient. Data in the supplemental data set shows that benthic crustaceans have moderate to high sensitivity to malathion exposure (LC_{50}s range from 0.5 to 290 μg/L for seven benthic species), and without a high-quality study for this important missing datum EPA criteria were not generated for malathion.

Using the log-triangular calculation (following the USEPA 1985 guidelines) and the acute chlorpyrifos and diazinon data sets, the following acute criteria were calculated. (Note: USEPA methodology uses *genus* mean acute values while *species* mean acute values are used in the UCDM. Since there is only one species from each genus in Tables 3 and 5, the final data sets would be the same in both schemes.)

	Example acute criterion = Final acute value/2
Chlorpyrifos:	Example final acute value (fifth percentile) = 0.052 μg/L
	Example acute criterion = 0.026 μg/L
Diazinon :	Example final acute value (fifth percentile) = 0.1662 μg/L
	Example acute criterion = 0.083 μg/L

According to the USEPA (1985) method, the criteria were rounded to two significant digits. The chlorpyrifos example acute criterion is higher than the acute criterion calculated by the UCDM (0.01 μg/L) by a factor of 2.6. The diazinon example acute criterion is lower than the acute criterion calculated using the Burr Type III distribution of the UCDM (0.2 μg/L) by approximately a factor of 2.

For the chronic criterion, there are only chlorpyrifos data for three species and the diazinon data set only has four species, which are not enough for the use of an SSD according to either method. The USEPA (1985) methodology contains a similar ACR procedure as the UCDM, to be used when three acceptable ACRs are available. The same three ACRs calculated for the UCDM (Tables 9 and 10) were calculated according to the USEPA (1985) methodology to give a final chlorpyrifos ACR of 2.2 and a final diazinon ACR of 2.3. Chronic criteria are calculated by dividing the final acute value by the final ACR:

Example chronic criterion = Final acute value/Final ACR
Chlorpyrifos example chronic criterion = 0.024 μg/L
Diazinon example chronic criterion = 0.072 μg/L

The chlorpyrifos example chronic criterion is a factor of 2.4 higher than the one recommended by the UCDM. The diazinon example chronic criterion is very similar to the one recommended by the UCDM.

It is anticipated that criteria from the UCDM will be fairly similar to those derived by the USEPA method for chemicals that have larger data sets, since the criteria calculation procedures are similar for such compounds. Many of the novel aspects of the UCDM were added to enable criteria generation for compounds with limited data sets or to incorporate other factors that affect toxicity, such as how to account for mixtures in criteria compliance, which other criteria methodologies do not include.

17 Final Criteria Statements

- Chlorpyrifos: Aquatic life should not be affected unacceptably if the 4-day average concentration of chlorpyrifos does not exceed 0.01 μg/L (10 ng/L) more than once every 3 years on the average and if the 1-h average concentration does not exceed 0.01 μg/L (10 ng/L) more than once every 3 years on the average. Mixtures of chlorpyrifos and other OPs should be considered in an additive manner (see Sect. 7).
- Diazinon: Aquatic life should not be affected unacceptably if the 4-day average concentration of diazinon does not exceed 0.07 μg/L (70 ng/L) more than once every 3 years on the average and if the 1-h average concentration does not exceed 0.2 μg/L (200 ng/L) more than once every 3 years on the average. Mixtures of diazinon and other OPs should be considered in an additive manner (see Sect. 7).
- Malathion: Aquatic life should not be affected unacceptably if the 4-day average concentration of malathion does not exceed 0.028 μg/L more than once every 3 years on the average and if the 1-h average concentration does not exceed 0.17 μg/L more than once every 3 years on average. Mixtures of malathion and other OPs should be considered in an additive manner (see Sect. 7).

18 Summary

A new methodology for deriving freshwater aquatic life water quality criteria, developed by the University of California Davis, was used to derive criteria for three organophosphate insecticides. The UC Davis methodology resulted in similar criteria to other accepted methods, and incorporated new approaches that enable criteria generation in cases where the existing USEPA guidance cannot be used. Acute and chronic water quality criteria were derived for chlorpyrifos (10 and 10 ng/L, respectively), diazinon (200 and 70 ng/L, respectively), and malathion (170 and 28 ng/L, respectively). For acute criteria derivation, Burr Type III SSDs were fitted to the chlorpyrifos and diazinon acute toxicity data sets while an alternative assessment factor procedure was used for malathion because that acute data set did not contain adequate species diversity to use a distribution. ACRs were used to calculate chronic criteria because there was a dearth of chronic data in all cases, especially for malathion, for which there was a lack of paired acute and chronic invertebrate data. Another alternate procedure enabled calculation of the malathion chronic criterion by combining a default ratio with the experimentally derived ratios. A review of the diazinon chronic criterion found it to be underprotective of cladoceran species, so a more protective criterion was calculated using a lower distributional estimate. The acute and chronic data sets were assembled using a transparent and consistent system for judging the relevance and reliability of studies, and the individual study review notes are included. The resulting criteria are unique in that they were reviewed to ensure particular protection of sensitive and threatened and endangered species, and mixture toxicity is incorporated into criteria compliance for all three compounds.

For chlorpyrifos and diazinon, the UCDM generated criteria similar to the long-standing USEPA (1985) method, with less taxa requirements, a more statistically robust distribution, and the incorporation of new advances in risk assessment and ecotoxicology. According to the USEPA (1985) method, the data set gathered for malathion would not be sufficient to calculate criteria because it did not contain data for a benthic crustacean. Benthic crustacean data is also required to use a distributional calculation method by the UCDM, but when data is lacking the UCDM provides an alternate calculation method. The resulting criteria are associated with higher, unquantifiable uncertainty, but they are likely more accurate than values generated using static safety factors, which are currently common in risk assessment.

Acknowledgments We would like to thank the following reviewers: D. McClure (CRWQCB-CVR), J. Grover (CRWQCB-CVR), S. McMillan (CDFG), J. P. Knezovich (Lawrence Livermore National Laboratory), and X. Deng (CDPR). Funding for this project was provided by the California Regional Water Quality Control Board, Central Valley Region (CRWQCB-CVR). The contents of this document do not necessarily reflect the views and policies of the CRWQCB-CVR, nor does mention of trade names or commercial products constitute endorsement or recommendation for use. Contents do not necessarily reflect the views or policies of the USEPA, nor does mention of trade names or commercial products constitute endorsement or recommendation for use.

References

Allison DT, Hermanutz RO (1977) Toxicity of diazinon to brook trout and fathead minnows. Environmental Research Laboratory-Duluth, Office of Research and Development, U.S. Environmental Protection Agency, Duluth, MN. EPA-600/3-77-060.

Anderson TD, Lydy MJ (2002) Increased toxicity to invertebrates associated with a mixture of atrazine and organophosphate insecticides. Environ Toxicol Chem 21:1507–1514.

Anderson BS, Phillips BM, Hunt JW, Connor V, Richard N, Tjeerdema RS (2006) Identifying primary stressors impacting macroinvertebrates in the Salinas River (California, USA): Relative effects of pesticides and suspended particles. Environ Poll 141:402–408.

Ankley GT, Collyard SA (1995) Influence of piperonyl butoxide on the toxicity of organophosphate insecticides to 3 species of freshwater benthic invertebrates. Comp Biochem Phys C 110:149–155.

Arthur JW, Zischke JA, Allen KN, Hermanutz RO (1983) Effects of diazinon on macroinvertebrates and insect emergence in outdoor experimental channels. Aquat Toxicol 4:283–301.

Bailey HC, Draloi R, Elphick JR, Mulhall A-M, Hunt P, Tedmanson L, Lovell A (2000) Application of *Ceriodaphnia dubia* for whole effluent toxicity tests in the Hawkesbury-Nepean watershed, New South Wales, Australia: method development and validation. Environ Toxicol Chem 19:88–93.

Bailey HC, Elphick JR, Drassoi R, Lovell A (2001) Joint acute toxicity of diazinon and ammonia to *Ceriodaphnia dubia*. Environ Toxicol Chem 20:2877–2882.

Bailey HC, Miller JL, Miller MJ, Wiborg LC, Deanovic L, Shed T (1997) Joint acute toxicity of diazinon and chlorpyrifos to *Ceriodaphnia dubia*. Environ Toxicol Chem 16:2304–2308.

Banks KE, Turner PK, Wood SH, Matthews C (2005) Increased toxicity to *Ceriodaphnia dubia* in mixtures of atrazine and diazinon at environmentally realistic concentrations. Ecotox Environ Safe 60:28–36.

Banks KE, Wood SH, Matthews C, Thuesen KA (2003) Joint acute toxicity of diazinon and copper to *Ceriodaphnia dubia*. Environ Toxicol Chem 22:1562–1567.

Barton JM (1988) Pesticide assessment guidelines subdivision D: product chemistry requirements for the manufacturing-use product, malathion insecticide: section 63–2 to 63–21, physical chemistry characteristics. Malathion registration standard, American Cyanamid company, Princetone, NJ. EPA MRID 40944103 and 40944104.

Belden JB, Lydy MJ (2000) Impact of atrazine on organophosphate insecticide toxicity. Environ Toxicol Chem 19:2266–2274.

Belden JB, Lydy MJ (2006) Joint toxicity of chlorpyrifos and esfenvalerate to fathead minnows and midge larvae. Environ Toxicol Chem 25:623–629.

Beyers DW, Keefe TJ, Carlson CA (1994) Toxicity of carbaryl and malathion to two federally endangered fishes, as estimated by regression and ANOVA. Environ Toxicol Chem 13:101–107.

Blakemore G, Burgess D (1990) Chronic toxicity of cythion to *Daphnia magna* under flow-through test conditions. Malathion registration standard. Columbia (MO), USA: Report by Analytical Bio-Chemistry Laboratories, Inc. submitted to U.S. Environmental Protection Agency. EPA MRID 41718401.

Borthwick PW, Patrick JM, Middaugh DP (1985) Comparative acute sensitivities of early life stages of Atherinid fishes to chlorpyrifos and thiobencarb. Arch Environ Contam Toxicol 14:465–473.

Bowman J (1988) Acute flow through toxicity of chlorpyrifos to bluegill sunfish (*Lepomis macrochirus*): project ID 37189. Columbia, MO: Unpublished study prepared by Analytical Biochemistry Laboratories, Inc. submitted to U.S. Environmental Protection Agency. EPA MRID 40840904.

Bowman BT, Sans WW (1979) The aqueous solubility of twenty-seven insecticides and related compounds. J Environ Sci Health B 14:625–634.

Bowman BT, Sans WW (1983a) Determination of octanol-water partitioning coefficients (Kow) of 61 organo-phsphorus and carbamate insecticides and their relationship to respective water solubility (S) values. J Environ Sci Health B 18:667–683.
Bowman BT, Sans WW (1983b) Further water solubility determination of insecticidal compounds. J Environ Sci Health B 18:221–227.
Brandt OM, Fujimura RW, Finlayson BJ (1993) Use of *Neomysis mercedis* (Crustacea, Mysidacea) for estuarine toxicity tests. Trans Am Fish Soc 122:279–288.
Brock TCM, Crum SJH, Van Wijngaarden R, Budde BJ, Tijink J, Zuppelli A, Leeuwangh P. (1992a) Fate and effects of the insecticide Dursban® 4E in indoor *Elodea*-dominated and macrophyte-free fresh-water model-ecosystems. I. Fate and primary effects of the active ingredient chlorpyrifos. Arch Environ Contam Toxicol 23:69–84.
Brock TCM, Van Den Bogaert M, Bos AR, Van Breukelen SWF, Reiche R, Terwoert J, Suykerbuyk REM, Roijackers RMM (1992b) Fate and effects of the insecticide Dursban® 4E in indoor *Elodea*-dominated and macrophyte-free freshwater model-ecosystems. II. Secondary effects on community structure. Arch Environ Contam Toxicol 23:391–409.
Brock TCM, Vet J, Kerkhofs MJJ, Lijzen J, Van Zuilekom WJ, Gijlstra R (1993) Fate and effects of the insecticide Dursban(R) 4e in indoor Elodea-dominated and macrophyte-free freshwater model-ecosystems. 3. Aspects of ecosystem functioning. Arch Environ Contam Toxicol 25:160–169.
Brown R, Hugo J, Miller J, Harrington C (1997) Chlorpyrifos acute toxicity to the amphipod *Hyalella azteca*. Lab project No. 971095: 91/414 ANNEX I 8.3.4. 27p. Midland, MI: Unpublished study prepared by the Dow Chemical Co. submitted to U.S. Environmental Protection Agency. EPA MRID 44345601.
Brust HF (1964) A summary of chemical and physical properties of O,O-diethyl-(3,5,6-trichloro-2-pyridyl) phosphorothioate. Unpublished report. DowElanco, Indianapolis, IN.
Brust HF (1966) A summary of chemical and physical properties of Dursban. Down to Earth 22:21–22.
Budavari S, O'Neil MJ, Smith A (1996) The Merck index: an encyclopedia of chemicals, drugs, and biological. Merck, Whitehouse Station, NJ.
Budischak SA, Belden LK, Hopkins WA (2009) Relative toxicity of malathion to Trematode-infected and noninfected *Rana palustris* tadpoles. Arch Environ Contam Toxicol 56:123–128.
Burgess D (1988) Acute flow through toxicity of chlorpyrifos to *Daphnia magna*: final Report No. 37190. 158 p. Columbia, MO: Unpublished study prepared by Analytical Biochemistry Laboratories, Inc. submitted to U.S. Environmental Protection Agency. EPA MRID 40840902.
Call DJ (1993) Validation study of a protocol for testing the acute toxicity of pesticides to invertebrates using the apple snail (*Pomacea paludosa*). 57 p. Superior, WI: Unpublished study prepared by University of Wisconsin-Superior submitted to U.S. Environmental Protection Agency. Cooperative Agreement No. CR 819612–01.
CARB (2010) California Ambient Air Quality Standards. [cited 2010 September 28]. California Air Resources Board. Available from: www.arb.ca.gov/research/aaqs/aaqs2.pdf.
Carpenter M (1990) Determination of the photolysis rate of ^{14}C malathion in pH 4 aqueous solution. Malathion registration standard. Analytical Bio-Chemistry Laboratories, Inc., Columbia, MO. EPA MRID 42015201.
CCME (2008) Canadian water quality guidelines for the protection of aquatic life: Chlorpyrifos. In: Canadian environmental quality guidelines, 1999, Canadian Council of Ministers of the Environment, Winnipeg.
CDFG (1992a) Test No. 133, acute, chlorpyrifos, *Neomysis mercedis*. California Department of Fish and Game, Elk Grove, CA.
CDFG (1992b) Test No. 139, acute, chlorpyrifos, *Ceriodaphnia dubia*. California Department of Fish and Game, Elk Grove, CA.
CDFG (1992c) Test No. 142, acute, chlorpyrifos, *Neomysis mercedis*. California Department of Fish and Game, Elk Grove, CA.

CDFG (1992d) Test No. 143, acute, chlorpyrifos, *Neomysis mercedis*. California Department of Fish and Game, Elk Grove, CA.
CDFG (1992e) Test No. 150, acute, chlorpyrifos, *Ceriodaphnia dubia*. California Department of Fish and Game, Elk Grove, CA.
CDFG (1992f) Test No. 157. 96-h acute toxicity of diazinon to *Ceriodaphnia dubia*. California Department of Fish and Game, Elk Grove, CA.
CDFG (1992g) Test No. 163. 96-h acute toxicity of diazinon to *Ceriodaphnia dubia*. California Department of Fish and Game, Elk Grove, CA.
CDFG (1992h) Test No. 162. 96-h acute toxicity of diazinon to *Neomysis mercedis*, Aquatic Toxicity Laboratory. California Department of Fish and Game, Elk Grove, CA.
CDFG (1992i) Test No. 168. 96-h acute toxicity of diazinon to *Neomysis mercedis*. Aquatic Toxicology Laboratory. California Department of Fish and Game, Elk Grove, CA.
CDFG (1998a) Test No. 122. 96-h acute toxicity of diazinon to *Ceriodaphnia dubia*, Aquatic Toxicology Laboratory. California Department of Fish and Game, Elk Grove, CA.
CDFG (1998b) Test 132. 96-h toxicity of diazinon to *Physa sp*. Aquatic Toxicology Laboratory. California Department of Fish and Game, Elk Grove, CA.
CDFG (1999) Test No. 61, 7-day chronic, chlorpyrifos, *Ceriodaphnia dubia*. California Department of Fish and Game, Elk Grove, CA.
CDFG (2010a) State and federally listed endangered and threatened animals of California. California Natural Diversity Database. California Department of Fish and Game, Sacramento, CA. Available from: http://www.dfg.ca.gov/biogeodata/cnddb/pdfs/TEAnimals.pdf
CDFG (2010b) State and federally listed endangered, threatened, and rare plants of California. California Natural Diversity Database. California Department of Fish and Game, Sacramento, CA. Available from: http://www.dfg.ca.gov/biogeodata/cnddb/pdfs/TEPlants.pdf
CDWR (1995) Compilation of sediment & soil standards, criteria & guidelines. Quality assurance technical document 7. California Department of Water Resources, Sacramento, CA.
Chakrabarti A, Gennrich SM (1987) Vapor pressure of chlorpyrifos, unpublished report. DowElanco, Indianapolis, IN.
Charizopoulos E, Papadopoulou-Mourkidou E (1999) Occurrence of pesticides in rain of the Axios River Basin, Greece. Environ Sci Technol 33:2363–2368.
Cheminova (1988) Product chemistry – Fyfanon technical; #63 – Physical and chemical characteristic. Malathion registration standard. A/S Cheminova, Lenvig, Denmark. EPA MRID 40966603.
Cohle P (1989) Early life stage toxicity of cythion to rainbow trout (*Oncorhynchus mykiss*) in a flow-through system. Malathion registration standard. Analytical Bio-Chemistry Laboratories, Inc., Columbia, MO. EPA MRID 41422401.
Cooke CM, Shaw G, Lester JN, Collins CD (2004) Determination of solid–liquid partition coefficients (K-d) for diazinon, propetamphos and cis-permethrin: implications for sheep dip disposal. Sci Total Environ 329:197–213.
Crommentuijn T, Sijm D, de Bruijn J, van Leeuwen K, van de Plassche E (2000) Maximum permissible and negligible concentrations for some organic substances and pesticides. J Environ Manag 58:297–312.
CSIRO (2001) BurrliOZ, version 1.0.13. [cited 28 September 2010]. Commonwealth Scientific and Industrial Research Organization, Australia. Available from: http://www.cmis.csiro.au/Envir/burrlioz/.
Cuppen JGM, Gylstra R, Vanbeusekom S, Budde BJ, Brock TCM (1995) Effects of nutrient loading and insecticide application on the ecology of Elodea-dominated freshwater microcosms. 3. Responses of macroinvertebrate detritivores, breakdown of plant litter, and final conclusions. Archiv Fur Hydrobiologie 134:157–177.
Deneer JW, Budde BJ, Weijers A (1999) Variations in the lethal body burdens of organophosphorus compounds in the guppy. Chemosphere 38:1671–1683.

Denton D, Wheelock C, Murray S, Deanovic L, Hammock B, Hinton D (2003) Joint acute toxicity of esfenvalerate and diazinon to larval fathead minnows (*Pimephales promelas*). Environ Toxicol Chem 22:336–341.

Dilling WL, Lickly LC, Lickly TD, Murphy PG, McKeller RL (1984) Organic-photochemistry. 19. Quantum yields for O,O-diethyl O-(3,5,6-trichloro-2-pyridinyl) phosphorothioate (chlorpyrifos) and 3,5,6-trichloro-2-pyridinol in dilute aqueous solutions and their environmental phtotransformation rates. Environ Sci Technol 18:540–543.

Dortland RJ (1980) Toxicological evaluation of parathion and azinphosmethyl in freshwater model ecosystems (Versl. Landbouwk. Onderz., no. 898). Center for Agricultural Publishing and Documentation Agric Res Rep, Wageningen, The Netherlands. 112 p.

Downey JR (1987) Henry's law constant for chlorpyrifos, unpublished report. DowElanco, Indianapolis, IN.

Drummond JN (1986) Solubility of chlorpyrifos in various solvents. DowElanco, Indianapolis, IN.

Eaton JG (1970) Chronic malathion toxicity to bluegill (*Lepomis macrochirus* Rafinesque). Water Res 4:673–684.

Eaton J, Arthur J, Hermanutz RO, Kiefer R, Mueller L, Anderson R, Erickson RJ, Nordling B, Rogers J, Pritchard H (1985) Biological effects of continuous and intermittent dosing of outdoor experimental streams with chlorpyrifos. In: Bahner RC, Hansen DJ (eds) Aquatic Toxicology and Hazard Assessment: Eighth Symposium: ASTM STP 891. American Society for Testing and Materials, Philadelphia, PA. p 85–118.

El Arab AE, Attar A, Ballhorn L, Freitag D, Korte F (1990) Behavior of diazinon in a perch species. Chemosphere 21:193–199.

El-Merhibi A, Kumar A, Smeaton T (2004) Role of piperonyl butoxide in the toxicity of chlorpyrifos to *Ceriodaphnia dubia* and *Xenopus laevis*. Ecotox Environ Safe 57:202–212.

Faust SD, Gomaa HM (1972) Chemical hydrolysis of some organic phosphorus and carbamate pesticides in aquatic environments. Environ Lett 3:171.

Felsot A, Dahm PA (1979) Sorption of organophosphorus and carbamate insecticides by soil. J Agric Food Chem 27:557–563.

Fendinger NJ, Glotfelty DE (1988) A laboratory method for the experimental determination of air-water Henry's law constants for several pesticides. Environ Sci Technol 22:1289–1293.

Fendinger NJ, Glotfelty DE (1990) Henry law constants for selected pesticides, PAHs and PCBs. Environ Toxicol Chem 9:731–735.

Fendinger NJ, Glotfelty DE, Freeman HP (1989) Comparison of 2 experimental techniques for determining air-water Henry law constants. Environ Sci Technol 23:1528–1531.

Fernández-Casalderrey A, Ferrando MD, Andreu-Moliner E (1995) Chronic toxicity of diazinon to *Daphnia magna*: effects on survival, reproduction, and growth. Toxicol Environ Chem 49:25–32.

Forbis A, Leak T (1994) Uptake, depuration and bioconcentration of ^{14}C malathion by bluegill sunfish (*Lepomis macrochirus*) under flow-through test conditions. Malathion registration standard. Unpublished study prepared by Cheminova Agro A/S, Lenvig, Denmark, submitted to U.S. Environmental Protection Agency. EPA MRIDs 43106401 and 43106402.

Freed VH, Chiou CT, Schmedding DW (1979a) Degradation of Selected Organophosphate Pesticides in Water and Soil. J Agric Food Chem 27:706–708.

Freed VH, Schmedding D, Kohnert R, Haque R. 1979b. Physical-chemical properties of several organophosphates – Some implication in environmental and biological behavior. Pestic Biochem Physiol 10:203–211.

Fujimura R, Finlayson B, Chapman G (1991) Evaluation of acute and chronic toxicity tests with larval striped bass. In: Mayes MA, Barron MG (eds) Aquatic Toxicology and Risk Assessment, 14th volume, STP 1124. American Society for Testing and Materials, Philadelphia, PA. p 193–211.

Fulton MH, Key PB (2001) Acetylcholinesterase inhibition in estuarine fish and invertebrates as an indicator of organophosphorus insecticide exposure and effects. Environ Toxicol Chem 20:37–45.

Geiger DL, Call DJ, Brooke LT (1984) Acute toxicities of organic chemicals to fathead minnows (*Pimephales promelas*). Lake Superior Research Institute, Superior, WI.

Geiger DL, Call DJ, Brooke LT (eds) (1988) Acute toxicities of organic chemicals to fathead minnows (*Pimephales promelas*), Volume IV. Center for Lake Superior Environmental Studies, University of Wisconsin-Superior, Superior, WI. 355 p.

Giddings JM, Biever RC, Annuziato MF, Hosmer AJ (1996) Effects of diazinon on large outdoor pond microcosms. Environ Toxicol Chem 15:618–629.

Giddings JM, Biever RC, Racke KD (1997) Fate of chlorpyrifos in outdoor pond microcosms and effects on growth and survival of bluegill sunfish. Environ Toxicol Chem 16:2353–2362.

Gilliom RJ (2007) Pesticides in U.S. streams and groundwater. Environ Sci Technol 41:3407–3413.

Glotfelty DE, Majewski MS, Seiber JN (1990) Distribution of several organophosphorus insecticides and their oxygen analogs in a foggy atmosphere. Environ Sci Technol 24:353–357.

Gomaa HM, Suffet IH, Faust SD (1969) Kinetics of hydrolysis of diazinon and diazoxon. Residue Rev 29:171–190.

Goodman LR, Hansen DJ, Cripe GM, Middaugh DP, Moore JC (1985) A new early life-stage toxicity test using the California grunion (*Leuresthes tenuis*) and results with chlorpyrifos. Ecotox Environ Safe 10:12–21.

Hall LW Jr, Anderson RD (2005) Acute toxicity of diazinon to the amphipod, *Gammarus pseudolimnaeus*: implications for water quality criteria. Bull Environ Contam Toxicol 74:94–99.

Harmon SM, Specht WL, Chandler GT (2003) A comparison of the daphnids *Ceriodaphnia dubia* and *Daphnia ambigua* for their utilization in routine toxicity testing in the Southeastern United States. Arch Environ Contam Toxicol 45:79–85.

Hartley GS, Graham-Bryce IJ (1980) Physical principles of pesticide behaviour (Vol. 1). Academic Press Inc., London, England.

Hermanutz RO (1978) Endrin and malathion toxicity to flagfish (*Jordanella floridae*). Arch Environ Contam Toxicol 7:159–168.

Hermanutz RO, Eaton JG, Mueller LH (1985) Toxicity of endrin and malathion mixtures to flagfish (*Jordanella floridae*). Arch Environ Contam Toxicol 14:307–314.

Hinckley DA, Bidleman TF, Foreman WT, Tuschall JR (1990) Determination of vapor-pressures for nonpolar and semipolar organic-compounds from gas-chromatographic retention data. J Chem Eng Data 35:232–237.

Holcombe GW, Phipps GL, Tanner DK (1982) The acute toxicity of kelthane, Dursban, disulfoton, pydrin, and permethrin to fathead minnows *Pimephales promelas* and rainbow trout *Salmo gairdneri*. Environ Pollut A 29:167–178.

Howard PH (1989) Handbook of environmental fate and exposure data for organic chemicals. Lewis Publishers, Chelsea, MI.

Hughes JS (1988) Toxicity of diazinon technical to *Selenastrum capricornutum,* Lab Sty N. 0267 – 40-1100-1. Unpublished study prepared by CIBA-GEIGY submitted to U.S. Environmental Protection Agency. EPA MRID 40509806.

Hummel RA, Crummet WB (1964) Solubility of ethyl O,O-diethyl O-(3,5,6-trichloro-2-pyridyl) phosphorothioate in various solvents. DowElanco, Indianapolis, IN.

Humphrey C, Klumpp DW (2003) Toxicity of chlorpyrifos to early life history stages of eastern rainbowfish *Melanotaenia splendida splendida* (Peters 1866) in tropical Australia. Environ Toxicol 18:418–427.

Hunt JW, Anderson BS, Phillips BM, Nicely PN, Tjeerdema RS, Puckett HM, Stephenson M, Worcester K, De Vlaming V (2003) Ambient toxicity due to chlorpyrifos and diazinon in a central California coastal watershed. Environ Monit Assess 82:83–112.

Hyder AH, Overmyer JP, Noblet R (2004) Influence of developmental stage on susceptibilities and sensitivities of *Simulium vittatum* IS-7 and *Simulium vittatum* IIIL-1 (Diptera: Simuliidae) to chlorpyrifos. Environ Toxicol Chem 23:2856–2862.

Iglesias-Jimenez E, Poveda E, Sanchez-Martin MJ, Sanchez-Camazano M (1997) Effect of the nature of exogenous organic matter on pesticide sorption by the soil. Arch Environ Contam Toxicol 33: 117–124.
Ingersoll CG, MacDonald DD, Wang N, Crane JL, Field LJ, Haverland PS, Kemble NE, Lindskoog RA, Severn C, Smorong DE (2000) Prediction of sediment toxicity using consensus-based freshwater sediment quality guidelines. EPA 905/R-00/007. Available from: http://www.cerc.usgs.gov/pubs/center/pdfdocs/91126.pdf.
Jarvinen AW, Nordling BR, Henry ME (1983) Chronic toxicity of Dursban (chlorpyrifos) to the fathead minnow (*Pimephales promelas*) and the resultant acetylcholinesterase inhibition. Ecotox Environ Safe 7:423–434.
Jarvinen AW, Tanner DK (1982) Toxicity of selected controlled release and corresponding unformulated technical grade pesticides to the fathead minnow *Pimephales promelas*. Environ Pollut A 27:179–195.
Jensen LD, Gaufin AR (1964a) Effects of ten organic insecticides on two species of stonefly naiads. Trans Am Fish Soc 93:27–34.
Jensen LD, Gaufin AR (1964b) Long term effects of organic insecticides on two species of stonefly naiads. Trans Am Fish Soc 93:357–363.
Jin-Clark Y, Lydy MJ, Zhu KY (2002) Effects of atrazine and cyanazine on chlorpyrifos toxicity in *Chironomus tentans* (Diptera: Chironomidae). Environ Toxicol Chem 21:598–603.
JMP (2004) Statistical discovery software, version 5.1.2. SAS Institute, Inc., Cary, NC.
Johnson WW, Finley MT (1980) Handbook of acute toxicity of chemicals to fish and aquatic invertebrates: summaries of toxicity tests conducted at Columbia National Fisheries Research Laboratory, 1965–78. U.S. Dept. of the Interior, Fish and Wildlife Service, Washington, DC. EPA MRID 40094602.
Kabler K (1989) Determination of aqueous solubility of ^{14}C malathion in pure water. Malathion registration standard. Analytical Bio-Chemistry laboratories, Inc., Columbia, MO. EPA MRID 41126201.
Kamiya M, Kameyama K (1998) Photochemical effects of humic substances on the degradation of organophosphorus pesticides. Chemosphere 36:2337–2344.
Kamrin MA, Montgomery JH (2000) Agrochemical and pesticide desk reference. Chapman & Hall/CRCnetBASE, Boca Raton, FL.
Kanazawa J (1975) Uptake and excretion of organophosphorus and carbamate insecticides by freshwater fish, motsugo, *Pseudorasbora parva*. Bull Environ Contam Toxicol 14:346–352.
Kanazawa J (1978) Bioconcentration ratio of diazinon by freshwater fish and snail. Bull Environ Contam Toxicol 20:613–617.
Kanazawa J (1981) Measurement of the bioconcentration factors of pesticides by fresh-water fish and their correlation with physiochemical properties or acute toxicities. Pestic Sci 12:417–424.
Kanazawa J (1989) Relationship between the soil sorption constants for pesticides and their physicochemical properties. Environ Toxicol Chem 8:477–484.
Karickhoff SW (1981) Semiempirical estimation of sorption of hydrophobic pollutants on natural sediments and soils. Chemosphere 10:833–846.
Keizer J, Dagostino G, Nagel R, Gramenzi F, Vittozzi L (1993) Comparative diazinon toxicity in guppy and zebra fish - different role of oxidative-metabolism. Environ Toxicol Chem 12:1243–1250.
Keizer J, Dagostino G, Vittozzi L (1991) The importance of biotransformation in the toxicity of xenobiotics to fish. 1. Toxicity and bioaccumulation of diazinon in guppy (*Poecilia reticulata*) and zebra fish (*Brachydanio rerio*). Aquat Toxicol 21:239–254.
Keller AE, Ruessler DS (1997) The toxicity of malathion to unionid mussels: Relationship to expected environmental concentrations. Environ Toxicol Chem 16:1028–1033.
Kennedy HD, Walsh DF (1970) Effects of malathion on two warmwater fishes and aquatic invertebrates in ponds. U.S. Bureau of sport fisheries and wildlife technical papers 55:3–13.

Kersting K, Van Den Brink PJ (1997) Effects of the insecticide Dursban(R)4E (active ingredient chlorpyrifos) in outdoor experimental ditches: Responses of ecosystem metabolism. Environ Toxicol Chem 16:251–259.

Kersting K, Van Wijngaarden R (1992) Effects of chlorpyrifos on a microecosystem. Environ Toxicol Chem 11:365–372.

Kidd H, James DR (1991) The Agrochemicals handbook. Royal Society of Chemistry. Cambridge, England.

Kikuchi M, Sasaki Y, Wakabayashi M (2000) Screening of organophosphate insecticide pollution in water by using *Daphnia magna*. Ecotox Environ Safe 47:239–245.

Kim YH, Woodrow JE, Seiber JN (1984) Evaluation of a gas-chromatographic method for calculating vapor-pressures with organo-phosphorus pesticides. J Chromatog 314:37–53.

Laetz CA, Baldwin DH, Collier TK, Hebert V, Stark JD, Scholz NL (2009) The synergistic toxicity of pesticide mixtures: Implications for risk assessment and the conservation of endangered Pacific salmon. Environ Health Persp 117:348–353.

Lahr J, Badji A, Marquenie S, Schuiling E, Ndour KB, Diallo AO, Everts JW (2001) Acute toxicity of locust insecticides to two indigenous invertebrates from Sahelian temporary ponds. Ecotox Environ Safe 48:66–75.

Landrum PF, Fisher SW, Hwang H (1999) Hazard evaluation of ten organophosphorus insecticides against the midge, *Chironomus riparius* via QSAR. SAR QSAR Environ Res 10:423–450.

Lartiges SB, Garrigues PP (1995) Degradation kinetics of organophosphorus and organonitrogen pesticides in different waters under various environmental-conditions. Environ Sci Technol 29:1246–1254.

Lide DR (2004) CRC handbook of chemistry and physics: A ready-reference book of chemical and physical data. CRC press, Boca Raton, FL.

Lien NTH, Adriaens D, Janssen CR (1997) Morphological abnormalities in African catfish (*Clarias gariepinus*) larvae exposed to malathion. Chemosphere 35:1475–1486.

Lydy MJ, Austin KR (2005) Toxicity assessment of pesticide mixtures typical of the Sacramento-San Joaquin Delta using *Chironomus tentans*. Arch Environ Contam Toxicol 48:49–55.

Lydy MJ, Belden JB, Ternes MA (1999) Effects of temperature on the toxicity of M-parathion, chlorpyrifos, and pentachlorobenzene to *Chironomus tentans*. Arch Environ Contam Toxicol 37:542–547.

Macalady DL, Wolfe NL (1983) New perspectives on the hydrolytic degradation of the organophosphorothioate insecticide chlorpyrifos. J Agr Food Chem 31:1139–1147.

Macek KJ (1975) Acute toxicity of pesticide mixtures to bluegills. Bull Environ Contam Toxicol 14:648–652.

Macek KJ, Hogan JW, Holz DD, Walsh DF (1972) Toxicity of insecticide Dursban to fish and aquatic invertebrates in ponds. Trans Am Fish Soc 101:420–427.

Macek KJ, Hutchinson C, Cope OB (1969) The effects of temperature on the susceptibility of bluegills and rainbow trout to selected pesticides. Bull Environ Contam Toxicol 4:174–183.

Mackay D, Shiu WY, Ma KC, Lee SC (2006) Handbook of Physical-Chemical Properties and Environmental Fate for Organic Chemicals, 2nd edn. CRC Press, Boca Raton, FL.

Mahar AM, Watzin MC (2005) Effects of metal and organophosphate mixtures on *Ceriodaphnia dubia* survival and reproduction. Environ Toxicol Chem 24:1579–1586.

Mansour M, Feicht EA, Behechti A, Schramm KW, Kettrup A (1999) Determination photostability of selected agrochemicals in water and soil. Chemosphere 39:575–585.

Martin H, Worthing CR (ed) (1977) The Pesticide Manual, 5th edition, a World Compendium. The British Crop Protection Council, Thornton Heath, UK.

Martinez-Tabche L, Galar MM, Olvera HE, Chehue RA, Lopez EL, Gomez-Olivan L, Sierra OT (2002) Toxic effect and bioavailability of malathion spiked in natural sediments from the Ignacio Ramirez dam on the snail *Stagnicola sp*. Ecotox Environ Safe 52:232–237.

Maul JD, Farris JL, Lydy MJ (2006) Interaction of chemical cues from fish tissues and organophosphorous pesticides on *Ceriodaphnia dubia* survival. Environ Pollut 141:90–97.

Mayer FL Jr, Ellersieck MR (1986) Manual of acute toxicity: interpretation and data base for 410 chemicals and 66 species of freshwater animals. Resource Publication No. 160. U.S. Fish and Wildlife Service, Washington, DC. EPAMRID 40098001.

Mayes M, Weinberg J, Rick D, Martin MD (1993) Chlorpyrifos: A life-cycle toxicity test with the fathead minnow, *Pimephales promelas* Rafinesque. Laboratory study number ES-DR-0043-4946-9, study ID: DECO-ES-2557B. Unpublished study prepared by the Dow Chemical Company, Midland, MI, submitted to the U.S. Environmental Protection Agency. EPA MRID 42834401.

McConnell LL, Lenoir JS, Datta S, Seiber JN (1998) Wet deposition of current-use pesticides in the Sierra Nevada mountain range, California, USA. Environ Toxicol Chem 17:1908–1916.

McDonald G, Karris GC, Chakrabarti A (1985) The melting behaviour, heat of melting, specific heat capacity, thermal conductivity, and vapor pressure of a recrystallized Dursban insecticide, unpublished report. DowElanco, Indianapolis, IN.

Medina D, Prieto A, Ettiene G, Buscema I, de V AA (1999) Persistence of organophosphorus pesticide residues in Limon River waters. Bull Environ Contam Toxicol 63:39–44.

Meikle RW, Kurihara NH, Devries DH (1983) Chlorpyrifos – the photo-decomposition rates in dilute aqueous-solution and on a surface, and the volatilization rate from a surface. Arch Environ Contam Toxicol 12:189–193.

Meikle RW, Youngson CR (1978) Hydrolysis rate of chlorpyrifos, O-O-diethyl O-(3,5,6-trichloro-2-pyridyl) phosphorothioate, and its dimethyl analog, chlorpyrifos-methyl, in dilute aqueous-solution. Arch Environ Contam Toxicol 7:13–22.

Melnikov NN (1971) Chemistry of pesticides. Residue Rev 36:1–480.

Moore MT, Lizotte Jr RE, Smith Jr S (2007) Toxicity evaluation of diazinon contaminated leaf litter. Bull Environ Contam Toxicol 78:158–161.

Nelson SM, Roline RA (1998) Evaluation of the sensitivity of rapid toxicity tests relative to daphnid acute lethality tests. Bull Environ Contam Toxicol 60:292–299.

Nguyen LTH, Janssen CR (2002) Embryo-larval toxicity tests with the African catfish (*Clarias gariepinus*): Comparative sensitivity of endpoints. Arch Environ Contam Toxicol 42:256–262.

NOAA (1999) Sediment quality guidelines developed for the National Status and Trends Program. [cited 2010 October 1]. National Atmospheric Association. Available from: http://response.restoration.noaa.gov/book_shelf/121_sedi_qual_guide.pdf.

Noblet JA, Smith LA, Suffet IH (1996) Influence of natural dissolved organic matter, temperature, and mixing on the abiotic hydrolysis of triazine and organophosphate pesticides. J Agric Food Chem 44:3685–3693.

Norberg-King TJ (1987) Toxicity data on diazinon, aniline, 2,4-Dimethylphenol. Memorandum to Stephan C, U.S. EPA, Duluth, MN and to Call D and Brooke L, Center for Lake Superior Environmental Studies, Superior, WI. U.S. Environmental Protection Agency, Duluth, MN.

Olvera-Hernandez E, Martinez-Tabche L, Martinez-Jeronimo F (2004) Bioavailability and effects of malathion in artificial sediments on *Simocephalus vetulus* (Cladocera, Daphniidae). Bull Environ Contam Toxicol 73:197–204.

Overmyer JP, Armbrust KL, Noblet R (2003) Susceptibility of black fly larvae (Diptera: Simuliidae) to lawn-care insecticides individually and as mixtures. Environ Toxicol Chem 22:1582–1588.

Palacio JA, Henao B, Velez JH, Gonzalez J, Parra CM (2002) Acute toxicity and bioaccumulation of pesticide diazinon in red tilapia (*Oreochromis niloticus* x *Mossambicus albina*). Environ Toxicol 17:334–340.

Pape-Lindstrom PA, Lydy MJ (1997) Synergistic toxicity of atrazine and organophosphate insecticides contravenes the response addition mixture model. Environ Toxicol Chem 16:2415–2420.

Patra RW, Chapman JC, Lim RP, Gehrke PC (2007) The effects of three organic chemicals on the upper thermal tolerances of four freshwater fishes. Environ Toxicol Chem 26:1454–1459.

Phillips TA, Summerfelt RC, Wu J, Laird DA (2003) Toxicity of chlorpyrifos adsorbed on humic colloids to larval walleye (*Stizostedion vitreum*). Arch Environ Contam Toxicol 45:258–263.

Phipps GL, Holcombe GW (1985) A method for aquatic multiple species toxicant testing – Acute toxicity of 10 chemicals to 5 vertebrates and 2 invertebrates. Environ Pollut A 38:141–157.

Post G, Schroeder T (1971) The toxicity of four insecticides to four Salmonid species. Bull Environ Contam Toxicol 6:144–155.

Pusey BJ, Arthington AH, McLean J (1994) Effects of a pulsed application of chlorpyrifos on macroinvertebrate communities in an outdoor artificial stream system. Ecotox Environ Safe 27:221–250.

Racke KD (1993) Environmental fate of chlorpyrifos. Rev Environ Contam Toxicol 131:1–150.

Raimondo S, Vivian DN, Barron MG (2010) Web-based Interspecies Correlation Estimation (Web-ICE) for Acute Toxicity: User Manual. Version 3.1. Office of Research and Development, U.S. Environmental Protection Agency, Gulf Breeze, FL. EPA/600/R-10/004.

Rawash IA, Gaaboub IA, El-Gayar FM, El-Shazli AY (1975) Standard curves for nuvacron, malathion, sevin, DDT, kelthane tested against the mosquito *Culex pipiens* L. and the microcrustacean *Daphnia magna* Straus. Toxicology 4:133–144.

Rawn GP, Webster GRB, Findlay GM (1978) Effect of pool bottom substrate on residues and bioactivity of chlorpyrifos, against larvae of *Culex tarsalis* (Diptera-Culicidae). Can Entomol 110:1269–1276.

Relyea RA (2005) The impact of insecticides and herbicides on the biodiversity and productivity of aquatic communities. Ecol Appl 15:618–627.

Rider CV, LeBlanc GA (2005) An integrated addition and interaction model for assessing toxicity of chemical mixtures. Toxicol Sci 87:520–528.

Rigertink RH, Kenaga EE (1966) Synthesis and insecticidal activity of some O,O-dialkyl O-3,5,6-trihalo-2-pyridyl phosphate and phosphorothioates. J Agric Food Chem 14:304–307.

Sabljic A, Gusten H, Verhaar H, Hermens J (1995) QSAR Modeling of soil sorption – Improvements and systematics of log K_{OC} vs log K_{OW} correlations. Chemosphere 31:4489–4514.

Sánchez M, Ferrando MD, Sancho E, Andreu-Moliner E (1998) Evaluation of a *Daphnia magna* renewal life-cycle test method with diazinon. J Environ Sci Health B 33:785–797.

Sánchez M, Ferrando MD, Sancho E, Andreu E (2000) Physiological perturbations in several generations *of Daphnia magna* Straus exposed to diazinon. Ecotox Environ Safe 46:87–94.

Sancho E, Ferrando MD, Andreu E, and Gamon M (1993) Bioconcentration and excretion of diazinon by eel. Bull Environ Contam Toxicol 50:578–585.

Sangster Research Laboratories (2004) LOGKOW. A databank of evaluated octanol-water partition coefficients (Log P). [cited 2010 September 28]. Available from: http://logkow.cisti.nrc.ca/logkow/.

Scharf J, Wiesiollek R, Bächmann K (1992) Pesticides in the atmosphere. Fresnius J Anal Chem 342:813–816.

Siefert R (1984) Effects of Dursban (chlorpyrifos) on non-target organisms in a natural pond undergoing mosquito control treatment. Unpublished report prepared by Environmental Research Laboratory- Duluth, Duluth, MN, submitted to the U.S. Environmental Protection Agency. EPA MRID 459401–096.

Siepmann S, Finlayson B (2000) Water quality criteria for diazinon and chlorpyrifos. California Department of Fish and Game, Rancho Cordova, CA. Administrative report 00–3.

Solomon HM, Weis JS (1979) Abnormal circulatory development in medaka caused by the insecticides carbaryl, malathion and parathion. Teratology 19:51–62.

Spieszalski WW, Niemczyk HD, Shetlar DJ (1994) Sorption of chlorpyrifos and fonofos on 4 soils and turfgrass thatch using membrane filters. J Environ Sci Heal B 29:1117–1136.

Steinberg CEW, Xu Y, Lee SK, Freitag D, Kettrup A (1993) Effect of dissolved humic material (DHM) on bioavailability of some organic xenobiotics to *Daphnia magna*. Chem Spec Bioavailab 5:1–9.

Surprenant DC (1988) The chronic toxicity of ^{14}C-diazinon technical to *Daphnia magna* under flow-through conditions, EPA guidelines No. 72-4. Unpublished study prepared by Springborn

Life Sciences, Inc., Wareham, MA, submitted to the U.S. Environmental Protection Agency. EPA MRID 40782302.

Sved D, Drottar KR, Swigert J, Smith GJ (1993) Chlorpyrifos: A flow-through life-cycle toxicity test with the saltwater mysid (*Mysidopsis bahia*): Final report. Lab project number: 103A-103 C. Dow contract study ID: ES-DR-0043-4946, ES-2506. 57 p. Unpublished study prepared by Wildlife International Ltd., Easton, MD, submitted to the U.S. Environmental Protection Agency. EPA MRID 42664901.

Teeter D (1988) Malathion (AC 6, 601): Hydrolysis. Malathion registration standard. American Cyanamid Company, Princeton, NJ, 1–64. EPA MRID 40941201.

TenBrook PL, Palumbo AJ, Fojut TL, Hann P, Karkoski J, Tjeerdema RS (2010) The University of California-Davis Methodology for deriving aquatic life pesticide water quality criteria. Rev Environ Contam Toxicol 209:1–155.

TenBrook PL, Tjeerdema RS, Hann P, Karkoski J (2009) Methods for deriving pesticide aquatic life criteria. Rev Environ Contam Toxicol 199:19–109.

Tietze NS, Hester PG, Hallmon CF, Olson MA, Shaffer KR (1991) Acute toxicity of mosquitocidal compounds to young mosquitofish, *Gambusia affinis*. J Am Mosq Cont Assoc 7:290–293.

Tomlin CDS (ed) 2003. The Pesticide Manual, A World Compendium, 13th edition. British Crop Protection Council, Alton, Hampshire, UK.

Tondreau R (1987) Malathion (AC 6,601): The determination of water solubility. Malathion registration standard. American Cyanamid Company, Princeton, NJ.

Tsuda T, Aoki S, Inoue T, Kojima M (1995) Accumulation and excretion of diazinon, fenthion and fenitrothion by killifish - Comparison of individual and mixed pesticides. Water Res 29:455–458.

Tsuda T, Aoki S, Kojima M, Harada H (1989) Bioconcentration and excretion of diazinon, IBP, malathion and fenitrothion by willow shiner. Toxicol Environ Chem 24:185–190.

Tsuda T, Aoki S, Kojima M, Harada H (1990) Bioconcentration and excretion of diazinon, IBP, malathion and fenitrothion by carp. Comp Biochem Physiol C 96:23–26.

Tsuda T, Kojima M, Harada H, Nakajima A, Aoki S (1997) Relationships of bioconcentration factors of organophosphate pesticides among species of fish. Comp Biochem Physiol C 116:213–218.

UKTAG (2008) Proposals for Environmental Quality Standards for Annex VIII Substances. UK Technical Advisory Group on the Water Framework Directive. Available at: http://www.wfduk.org/stakeholder_reviews/stakeholder_review_1-2007/LibraryPublicDocs/final_specific_pollutants.

USEPA (1985) Guidelines for deriving numerical national water quality criteria for the protection of aquatic organisms and their uses. U.S. Environmental Protection Agency, Springfield, VA. PB-85-227049.

USEPA (1986a) Ambient water quality criteria for chlorpyrifos. U.S. Environmental Protection Agency, Washington, DC. EPA 440/5-86-005.

USEPA (1986b) Quality criteria for water. U.S. Environmental Protection Agency, Washington, DC. EPA 440/5-86-001.

USEPA (2002) Interim reregistration eligibility decision for chlorpyrifos. U.S. Environmental Protection Agency, Washington, DC. EPA 738-R-01-007.

USEPA (2004a) Interim reregistration eligibility decision, diazinon. U.S. Environmental Protection Agency, Washington, DC. EPA 738-R-04-006.

USEPA (2004b) The incidence and severity of sediment contamination in surface waters of the United States: National sediment quality survey, 2nd edition. U.S. Environmental Protection Agency, Washington, DC. EPA-823-R-04-007. Available from: http://water.epa.gov/scitech/swguidance/waterquality/cs/pubs/upload/2004_12_09_cs_report_2004_nsqs2ed-complete.pdf.

USEPA (2005) Aquatic Life Ambient Water Quality Criteria. Diazinon. U.S. Environmental Protection Agency, Washington, DC. EPA-822-R-05-006.

USEPA (2009a) Sediment Quality Guidelines website. [cited 2010 October 1]. U.S. Environmental Protection Agency. Available from: http://water.epa.gov/scitech/swguidance/waterquality/cs/library_guidelines.cfm
USEPA (2009b) National Ambient Air Quality Standards website. [cited 2010 October 1]. U.S. Environmental Protection Agency. Available from: http://www.epa.gov/air/criteria.html.
USFDA (2000) Industry activities staff booklet. U.S. Food and Drug Administration, Washington, DC. Available from: http://www.cfsan.fda.gov/~lrd/fdaact.html.
USFWS (2010) Species Reports. Endangered Species Program. U.S. Fish and Wildlife Service. Available from: http://www.fws.gov/endangered/; http://ecos.fws.gov/tess_public/pub/listedAnimals.jsp; http://ecos.fws.gov/tess_public/pub/listedPlants.jsp.
Van Breukelen SWF, Brock TCM (1993) Response of a macroinvertebrate community to insecticide application in replicated freshwater microcosms with emphasis on the use of principal component analysis. Sci Total Environ Supplement: 1047–1058.
Van Den Brink PJ, Van Donk E, Gylstra R, Crum SJH, Brock TCM (1995) Effects of chronic low concentrations of the pesticides chlorpyrifos and atrazine in indoor freshwater microcosms. Chemosphere 31:3181–3200.
Van Den Brink PJ, Van Wijngaarden RPA, Lucassen WGH, Brock TCM, Leeuwangh P (1996) Effects of the insecticide Dursban(R) 4E (active ingredient chlorpyrifos) in outdoor experimental ditches. 2. Invertebrate community responses and recovery. Environ Toxicol Chem 15:1143–1153.
Van Der Geest HG, Greve GD, Boivin M-E, Kraak MHS, Gestel CAM (2000) Mixture toxicity of copper and diazinon to larvae of the mayfly (*Ephoron virgo*) judging additivity at different effect levels. Environ Toxicol Chem 19:2900–2905.
Van Der Hoeven N, Gerritsen AAM (1997) Effects of chlorpyrifos on individuals and populations of *Daphnia pulex* in the laboratory and field. Environ Toxicol Chem 16:2438–2447.
Van Donk E, Prins H, Voogd HM, Crum SJH, Brock TCM (1995) Effects of nutrient loading and insecticide application on the ecology of *Elodea*-dominated freshwater microcosms. 1. Responses of plankton and zooplanktivorous insects. Arch Hydrobiol 133:417–439.
Van Wijngaarden R, Leeuwangh P, Lucassen WGH, Romijn K, Ronday R, Vandervelde R, Willigenburg W (1993) Acute toxicity of chlorpyrifos to fish, a newt, and aquatic invertebrates. Bull Environ Contam Toxicol 51:716–723.
Van Wijngaarden RPA (1993) Comparison of response of the mayfly *Cloeon dipterum* to chlorpyrifos in a single species toxicity test, laboratory microcosms, outdoor ponds and experimental ditches. Sci Total Environ Supplement: 1037–1046.
Van Wijngaarden RPA, Brock TCM, Douglas MT (2005) Effects of chlorpyrifos in freshwater model ecosystems: the influence of experimental conditions on ecotoxicological thresholds. Pest Manag Sci 61:923–935.
Van Wijngaarden RPA, Leeuwangh V (1989) Relation between toxicity in laboratory and pond: an ecotoxicological study with chlorpyrifos. In: Proceedings of the International Symposium on Crop Protection (Mededelingen van de Faculteit Landbouwwetenschappen) volume 54(3b); 1989; Rijksuniversiteit Gent, Ghent, Belgium. p 1061–1069.
Van Wijngaarden RPA, Van Den Brink PJ, Crum SJH, Voshaar JHO, Brock TCM, Leeuwangh P (1996) Effects of the insecticide Dursban(R) 4E (active ingredient chlorpyrifos) in outdoor experimental ditches. 1. Comparison of short-term toxicity between the laboratory and the field. Environ Toxicol Chem 15:1133–1142.
Varó I, Serrano R, Pitarch E, Amat F, Lopez FJ, Navarro JC (2002) Bioaccumulation of chlorpyrifos through an experimental food chain: Study of protein HSP70 as biomarker of sublethal stress in fish. Arch Environ Contam Toxicol 42:229–235.
Venturino A, Gauna LE, Bergoc RM, Dedangelo AMP (1992) Effect of exogenously applied polyamines on malathion toxicity in the toad *Bufo arenarum hensel*. Arch Environ Contam Toxicol 22:135–139.
Verschueren K (1996) Handbook of environmental data on organic chemicals. Van Nostrand Reinhold, New York.

Ward S, Arthington AH, Pusey BJ (1995) The effects of a chronic application of chlorpyrifos on the macroinvertebrate fauna in an outdoor artificial stream system – species responses. Ecotox Environ Safe 30:2–23.

Werner I, Deanovic LA, Connor V, de Vlaming V, Bailey HC, Hinton DE (2000) Insecticide-caused toxicity to *Ceriodaphnia dubia* (Cladocera) in the Sacramento-San Joaquin River Delta, California, USA. Environ Toxicol Chem 19:215–227.

Wheelock CE, Eder KJ, Werner I, Huang HZ, Jones PD, Brammell BF, Elskus AA, Hammock BD (2005) Individual variability in esterase activity and CYP1A levels in Chinook salmon (*Oncorhynchus tshawyacha*) exposed to esfenvalerate and chlorpyrifos. Aquat Toxicol 74:172–192.

Wolfe NL, Zepp RG, Gordon JA, Baughman GL, Cline DM (1977) Kinetics of chemical degradation of malathion in water. Environ Sci Technol 11:88–93.

Worthing CR (ed) (1991) The Pesticide Manual, 9th Edition, A World Compendium. The British Crop Protection Council, Surrey, UK.

Wu L, Green RL, Liu G, Yates MV, Pacheco P, Gan J, Yates SR (2002) Partitioning and persistence of trichlorfon and chlorpyrifos in a creeping bentgrass putting green. J Environ Qual 31:889–895.

Zabik JM, Seiber JN (1993) Atmospheric transport of organophosphate pesticides from California's Central Valley to the Sierra Nevada mountains. J Environ Qual 22:80–90.

Aquatic Life Water Quality Criteria Derived via the UC Davis Method: II. Pyrethroid Insecticides

Tessa L. Fojut, Amanda J. Palumbo, and Ronald S. Tjeerdema

1 Introduction

Pyrethroid insecticides are broad spectrum agents that have been widely detected in sediments and surface waters in the USA (Amweg et al. 2006; Budd et al. 2007; Gan et al. 2005; Hladik and Kuivila 2009; Weston et al. 2004). They are hydrophobic compounds that primarily partition to sediments and solid materials in the water column, and exposure to pyrethroid-contaminated sediments has been demonstrated to produce toxicity in the environment (Anderson et al. 2006; Holmes et al. 2008; Phillips et al. 2010; Weston et al. 2004; Weston et al. 2005). Only very low concentrations are found freely dissolved in the aqueous phase, but these pesticides are still of concern to water quality managers because they exhibit toxicity to aquatic organisms at very low concentrations (<1 μg/L). Water quality regulators in the USA are required, under the Clean Water Act (section 303(c)(2)(B)), to provide numeric water quality criteria for priority pollutants that could reasonably be expected to interfere with the designated uses of a state's waters. Numeric water quality criteria are chemical concentrations in water bodies that should protect aquatic wildlife from the toxic effects of those chemicals, if these concentrations are not exceeded. Numeric criteria are derived using existing toxicity data; consequently, criteria calculation is dependent on the availability of these data. In the USA, there are currently no numeric criteria available for the pyrethroids, and many of the available pyrethroid data sets do not meet the requirements of the 1985 US Environmental Protection Agency (USEPA) criteria derivation methodology (USEPA 1985). One of the goals of developing the UC Davis methodology (UCDM) was to be able to derive criteria for compounds that do not meet all of the USEPA (1985) data requirements, such as the pyrethroids.

T.L. Fojut (✉) • A.J. Palumbo • R.S. Tjeerdema
Department of Environmental Toxicology, College of Agricultural and Environmental Sciences, University of California, Davis, CA 95616-8588, USA
e-mail: tlfojut@ucdavis.edu

R.S. Tjeerdema (ed.), *Aquatic Life Water Quality Criteria for Selected Pesticides*, Reviews of Environmental Contamination and Toxicology 216,
DOI 10.1007/978-1-4614-2260-0_2,

The UCDM is an updated water quality criteria derivation methodology that was designed to be more flexible than the USEPA (1985) methodology, and to incorporate the results from new research in environmental toxicology and risk assessment. Like the USEPA (1985) method, the UCDM continues to recommend the use of a species sensitivity distribution (SSD) for criteria calculation, and an acute-to-chronic ratio (ACR) when chronic data are limited. The main procedures of the UCDM that differ from those of the USEPA method are that the UCDM provides for a thorough and transparent study evaluation procedure, a more advanced SSD, alternate procedures if data requirements for the SSD or ACR cannot be met, and a consideration for the toxicity of chemical mixtures. Previous publications have described why there was a need for a new methodology (TenBrook et al. 2009), the rationale behind the development of this new methodology, and detailed instructions for UCDM criteria derivation (TenBrook et al. 2010).

This paper is the second in a series in which water quality criteria were derived for nine pesticides: chlorpyrifos, diazinon, malathion, bifenthrin, cyfluthrin, cypermethrin, λ-cyhalothrin, permethrin, and diuron. In this article, we describe the derivation of water quality criteria for five pyrethroid insecticides (bifenthrin, cyfluthrin, cypermethrin, λ-cyhalothrin, and permethrin) according to the UCDM; we have also extended this review to render it wide ranging and useful as a review of the current knowledge regarding the risk to aqueous ecosystems of the pyrethroids' toxicity.

2 Data Collection and Evaluation

Bifenthrin ((2-methyl[1,1′-biphenyl]-3-yl)methyl (1*R*,3*R*)-rel-3-[(1*Z*)-2-chloro-3,3,3-trifluoro-1-propenyl]-2,2-dimethylcyclopropanecarboxylate), cyfluthrin (cyano(4-fluoro-3-phenoxyphenyl)methyl 3-(2,2-dichloroethenyl)-2,2-dimethylcyclopropanecarboxylate (unstated stereochemistry)), cypermethrin (cyano(3-phenoxyphenyl) methyl 3-(2,2-dichloroethenyl)-2,2-dimethylcyclopropanecarboxylate), λ-cyhalothrin ([1α(*S**), 3α (*Z*)]-(±)-cyano-(3-phenoxyphenyl)methyl 3-(2-chloro-3,3,3-trifluoro-1-propenyl)-2,2-dimethylcyclopropanecarboxylate), and permethrin ((3-phenoxyphenyl) methyl 3-(2,2-dichloroethenyl)-2,2-dimethylcyclopropanecarboxylate) are widely applied pyrethroid insecticides. Bifenthrin and permethrin are type I pyrethroids while cyfluthrin, cypermethrin, and λ-cyhalothrin are type II pyrethroids (containing an α-cyano moiety); the two types are distinguished by slightly different toxicological mechanisms (Breckenridge et al. 2009). These pyrethroids are hydrophobic organic compounds that are moderately persistent (see Tables S1 and S2 of the Supporting Material http://extras.springer.com/). Based on their physical–chemical properties (Table 1), they are likely to partition to sediments from the aqueous phase, and are not likely to volatilize.

Aquatic toxicity effects studies were identified in the peer-reviewed open literature and from unpublished studies submitted to the USEPA and California Department of Pesticide Regulation (CDPR) for bifenthrin (~40), cyfluthrin (~53),

Table 1 Physical–chemical properties of five selected pyrethroids

	Bifenthrin	Cyfluthrin	Cypermethrin	λ-Cyhalothrin	Permethrin
Molecular weight	422.87	434.3	416.3	449.850	391.288
Density (g/mL)	1.21 (geomean, $n = 2$)	1.28[d] (20°C)	1.24[d] (20°C)	1.33 (25°C[d,f])	1.23 (geomean, $n = 2$)
Water solubility (mg/L)	0.001 (geomean, $n = 2$)	0.0023[e] (20°C)	0.004 (geomean, $n = 2$)	0.0047 (geomean, $n = 4$)	0.0057 (geomean, $n = 2$)
Melting point (°C)	69.3 (geomean, $n = 2$)	60[d]	71.2 (geomean of extremes)	48.3 (geomean of extremes)	36.4 (geomean of extremes)
Vapor pressure (Pa)	2.41×10^{-5} (geomean, $n = 2$)	2×10^{-6}[e]	2.87×10^{-7} (geomean, $n = 2$)	2.0×10^{-7} (20°C) (geomean, $n = 3$)	3.74×10^{-6} (geomean, $n = 4$)
Henry's law constant (K_H) (Pa m^3 mol^{-1})	0.24 (geomean, $n = 2$)	0.37[e]	0.0238 (geomean, $n = 3$)	1.96×10^{-2} (geomean, $n = 2$)	0.12 (geomean, $n = 2$)
Log K_{oc}[a]	5.29 (geomean, $n = 7$)	5.09[e] (mean, $n = 4$)	5.49[e] (mean, $n = 3$)	5.52 (geomean, $n = 2$)	5.12 (geomean, $n = 2$)
Log K_{ow}[b]	6.00[c]	5.97[e] (mean, $n = 4$)	6.57 (geomean, $n = 2$)	7.0[d,e,f]	6.3 (geomean, $n = 2$)

[a] Log-normalized organic carbon–water partition coefficient
[b] Log-normalized octanol–water partition coefficient
[c] Sangster Research Laboratories (2010)
[d] Tomlin (2003)
[e] Laskowski (2002)
[f] Mackay et al. (2006)

cypermethrin (~108), λ-cyhalothrin (~65), and permethrin (~155). Each study was reviewed according to the UCDM paradigm to determine the usefulness of these studies for criteria derivation. Studies were divided into three categories to be rated: (1) single-species effects, (2) ecosystem-level studies, and (3) terrestrial wildlife studies.

The UCDM provides a detailed numeric rating scheme for single-species effects studies that assigns (1) a relevance score and (2) a reliability score, which is summarized in the first chapter of this volume (Palumbo et al. (2012)). The possible relevance scores were relevant (R), less relevant (L), or not relevant (N). The studies rated N were deemed irrelevant for criteria derivation, and only the relevant (R) and less relevant (L) studies were evaluated for reliability. For all studies, study details and scoring were summarized in data summary sheets (Supporting Material http://extras.springer.com/). The reliability evaluation assigned possible scores of reliable (R), less reliable (L), or not reliable (N) so that each single-species study is described by a two-letter code, corresponding to the relevance and reliability ratings. The only studies used directly in criteria calculation were those rated as relevant and reliable (RR), which are summarized in Table 11. Studies that were rated as relevant and less reliable (RL), less relevant and reliable (LR), or less relevant and less reliable (LL) were used to evaluate the derived criteria against data for any particularly sensitive, threatened, or endangered species found in these data sets. Studies that were rated N for either relevance or reliability were not considered in any aspect of criteria derivation.

Multispecies studies conducted in mesocosms, microcosms, and other field and laboratory ecosystems were rated for reliability. The results of the studies that were rated reliable (R) or less reliable (L) were compared to the derived criteria to ensure that they are protective of ecosystems. Studies of the effects of pyrethroids on mallard ducks were rated for reliability using the terrestrial wildlife evaluation. Mallard studies rated as reliable (R) or less reliable (L) were used to consider bioaccumulation of pyrethroids.

3 Data Reduction

As described in Palumbo et al. (2012), multiple toxicity values for a given species in the acceptable data set were combined into one species mean acute value (SMAV) or one species mean chronic value (SMCV). Some data that were rated RR were excluded from the final data set for one or more of the following reasons: flow-through tests are preferred over static tests, a test with a more sensitive life stage of the same species was available, more appropriate exposure durations were available, and tests with more sensitive end points were available (Tables S3–S6, Supporting Material http://extras.springer.com/). For bifenthrin, the final acceptable data sets contain 8 SMAVs and 2 SMCVs (Tables 2 and 3), the final cyfluthrin data sets contain 8 SMAVs and 3 SMCVs (Tables 4 and 5), the final cypermethrin data sets contain 14 SMAVs and 1 SMCV (Tables 6 and 7), the final λ-cyhalothrin data sets contain 20 SMAVs and 2 SMCVs (Tables 8 and 9), and the final permethrin data sets contain 19 SMAVs and 3 SMCVs (Tables 10 and 11).

4 Acute Criterion Calculations

An acute data set must have species representing five taxa to use a SSD to calculate the acute criterion; the five taxa are a warm water fish, a species in the family Salmonidae, a planktonic crustacean, a benthic crustacean, and an insect. The final acute data sets for each of the five pyrethroids (Tables 2, 4, 6, 8, and 10) met the five taxa requirement. Log-logistic distributions were fit to the bifenthrin and cyfluthrin acute data sets using the ETX 1.3 software (Aldenberg 1993) because there were between five and eight SMAVs in each of these data sets. The Burr Type III distribution was fit to the acute λ-cyhalothrin and permethrin data sets because there were more than eight SMAVs in these data sets. Of the three related distributions in the Burr Type III SSD, the Burr III

Table 2 Final acute toxicity data set for bifenthrin

Species	Test type	Meas/ Nom	Chemical grade (%)	Duration (h)	Temp (°C)	End point	Age/size	LC/ EC_{50} (μg/L)	Reference
Ceriodaphnia dubia	SR	Est	97.8	96	24.0–24.7	Mortality	<24 h	0.078	Guy (2000a)
C. dubia	S	Nom	97.0	48	25	Mortality	<24 h	0.142	Wheelock et al. (2004)
Geometric mean								0.105	
Chironomus dilutes	FT	Nom	100	96	23 ± 1	Mortality	Third instar	2.615	Anderson et al. (2006)
Daphnia magna	FT	Nom	88.4	48	20–21	Mortality	<24 h	1.6	Surprenant (1983)
Hyalella azteca	S	Nom	100.0	96	23 ± 1	Mortality	7–14 days	0.0093	Anderson et al. (2006)
H. azteca	SR	Est	98	96	23 ± 1	Mortality	7–14 days	0.0027	Weston and Jackson (2009)
H. azteca	SR	Est	98	96	23 ± 1	Mortality	7–14 days	0.0073	Weston and Jackson (2009)
H. azteca	SR	Est	98	96	23 ± 1	Mortality	7–14 days	0.0080	Weston and Jackson (2009)
H. azteca	SR	Est	98	96	23 ± 1	Mortality	7–14 days	0.0082	Weston and Jackson (2009)
Geometric mean								0.0065	
Lepomis macrochirus	FT	Nom	88.4	96	21–22	Mortality	2.5 g, 8 mm	0.35	Hoberg (1983a)
Oncorhynchus mykiss	FT	Nom	88.4	96	11–12	Mortality	1.0 g, 46 mm	0.15	Hoberg (1983b)
Pimephales promelas	S	Meas	96.2	96	25 ± 1	Mortality	40 days, 0.059 g	0.21	McAllister (1988)
P. promelas	SR	Est	97.8	96	24.0–24.5	Mortality	8 days, 0.0039–0.0052 g	0.78	Guy (2000b)
Geometric mean								0.405	
Procloeon sp.	S	Nom	100.0	48	23 ± 1	Mortality	0.5–1.0 cm	0.0843	Anderson et al. (2006)

All studies were rated relevant and reliable (RR)

Est Toxicity values were calculated based on estimated concentrations (calculated from the recovery of some concentrations), *S* static, *SR* static renewal, *FT* flow through

Table 3 Final chronic toxicity data set for bifenthrin

Species	Test type	Meas/Nom	Chemical grade (%)	Duration (days)	Temp (°C)	End point	Age/size	NOEC (μg/L)	LOEC (μg/L)	MATC (μg/L)	Reference
Daphnia magna	FT	Meas	97.0	21	19–22	Reproduction	<24 h	0.0013	0.0029	0.0019	Burgess (1989)
Pimephales promelas	FT	Meas	96.2	92	25	Mortality	<48 h	0.040	0.090	0.060	McAllister (1988)

All studies were rated relevant and reliable (RR)
FT flow through

Table 4 Final acute toxicity data set for cyfluthrin

Species	Test type	Meas/Nom	Chemical grade (%)	Duration (h)	Temp (°C)	End point	Age/size	LC/EC$_{50}$ (μg/L) (95% CI)	Reference
Aedes aegypti Rockefellar	S	Nom	93.0	24	25	Mortality	Early fourth instar	1 (1–2)	Rodriguez et al. (2007)
A. aegypti Nicaragua	S	Nom	93.0	24	25	Mortality	Early fourth instar	0.5 (0.5–0.6)	Rodriguez et al. (2007)
A. aegypti Peru	S	Nom	93.0	24	25	Mortality	Early fourth instar	0.3 (0.1–0.4)	Rodriguez et al. (2007)
A. aegypti Geometric mean								0.5	
Ceriodaphnia dubia	S	Nom	97.0	48	25	Mortality	<24 h	0.344 ± 0.041	Wheelock et al. (2004)
C. dubia	S	Nom	99.0	96	21	Mortality	<24 h	0.093 (0.050–0.146)	Yang et al. (2007)
C. dubia	S	Nom	99.0	96	21	Mortality	<24 h	0.136 (0.103–0.185)	Yang et al. (2007)
C. dubia	S	Nom	99.0	96	21	Mortality	<24 h	0.189 (0.112–0.292)	Yang et al. (2007)
C. dubia	S	Nom	99.0	96	21	Mortality	<24 h	0.134 (0.097–0.194)	Yang et al. (2007)
C. dubia	S	Nom	99.0	96	21	Mortality	<24 h	0.170 (0.121–0.229)	Yang et al. (2007)
C. dubia	S	Nom	99.0	96	21	Mortality	<24 h	0.145 (0.105–0.185)	Yang et al. (2007)
C. dubia	S	Nom	99.0	96	21	Mortality	<24 h	0.102 (0.027–0.395)	Yang et al. (2007)
C. dubia	S	Nom	99.0	96	21	Mortality	<24 h	0.159 (0.105–0.234)	Yang et al. (2007)
C. dubia	S	Nom	99.0	96	21	Mortality	<24 h	0.180 (0.127–0.280)	Yang et al. (2007)
Geometric mean								0.155	
Daphnia magna	FT	Meas	98.6	48	19	Mortality	<24 h (first instar)	0.16 (0.14–0.18)	Burgess (1990)
Hyalella azteca	SR	Est	98.0	96	23	Mortality	7–14 days	0.0017 (0.0011–0.0023)	Weston and Jackson (2009)
H. azteca	SR	Est	98.0	96	23	Mortality	7–14 days	0.0023 (0.0009–0.0028)	Weston and Jackson (2009)

(continued)

Table 4 (continued)

Species	Test type	Meas/Nom	Chemical grade (%)	Duration (h)	Temp (°C)	End point	Age/size	LC/EC_{50} (μg/L) (95% CI)	Reference
H. azteca	SR	Est	98.0	96	23	Mortality	7–14 days	0.0031 (0.0021–0.0046)	Weston and Jackson (2009)
Geometric mean								0.0023	
Lepomis macrochirus	FT	Meas	97.6	96	22	Mortality	0.82 g, 31.8 mm	0.998	Gagliano (1994)
Oncorhynchus mykiss	FT	Meas	97.6	96	11	Mortality	0.92 g, 39 mm	0.209	Gagliano and Bowers (1994)
O. mykiss	FT	Meas	97.6	96	12	Mortality	1.4 g, 43.3 mm	0.302 (0.240–0.432)	Bowers (1994)
Geometric mean								0.2512	
Pimephales promelas	FT	Meas	99.0	96	25	Mortality	30-day old	2.49	Rhodes et al. (1990)
Procambarus clarkii	FT	Meas	97.0	96	20	Mortality	0.59 g, 29 mm	0.062	Surprenant (1990)

All studies were rated relevant and reliable (RR). *Est* Toxicity values were calculated based on estimated concentrations (calculated from the recovery of some concentrations)

S static, *SR* static renewal, *FT* flow through, *95% CI* 95% confidence interval

Table 5 Final chronic toxicity data set for cyfluthrin

Species	Test type	Meas/Nom	Chemical (%)	Duration (days)	Temp (°C)	End point	Age/size	NOEC (μg/L)	LOEC (μg/L)	MATC (μg/L)	Reference
Daphnia magna	FT	Meas	94.7	21	20	Reproduction (young/female/day)	<24 h	0.020	0.041	0.02864	Forbis et al. (1984)
D. magna	FT	Meas	94.7	21	20	Length	<24 h	0.020	0.041	0.02864	Forbis et al. (1984)
Geometric mean										0.02864	
Oncorhynchus mykiss	FT	Meas	96.0	58	9.4	Biomass/chamber	Eggs	0.01	0.0177	0.0133	Carlisle (1985)
O. mykiss	FT	Meas	96.0	58	9.4	Mean weight/fish	Eggs	0.01	0.0177	0.0133	Carlisle (1985)
Geometric mean										0.0133	
Pimephales promelas	FT	Meas	99.0	7–61	25	F_0 survival	Eggs	0.14	0.29	0.20	Rhodes et al. (1990)
P. promelas	FT	Meas	99.0	61–120	25	F_0 survival	Eggs	0.14	0.29	0.20	Rhodes et al. (1990)
P. promelas	FT	Meas	99.0	90	25	F_1 % hatch	Eggs	0.14	0.29	0.20	Rhodes et al. (1990)
P. promelas	FT	Meas	99.0	60	25	F_1 survival	Eggs	0.14	0.29	0.20	Rhodes et al. (1990)
Geometric mean										0.20	

All studies were rated relevant and reliable (RR)
S static, *SR* static renewal, *FT* flow through

Table 6 Final acute toxicity data set for cypermethin

Species	Test type	Meas/ Nom	Chemical grade (%)	Duration (h)	Temp (°C)	End point	Age/size	LC/EC_{50} (μg/L) (95% CI)	Reference
Aedes aegypti	S	Nom	>85	24	18	Mortality	Larvae	1 (0.4–4)	Stephenson (1982)
Asellus aquaticus	S	Nom	>85	24	15	Mortality	3–8 mm	0.2 (0.1–0.4)	Stephenson (1982)
Ceriodaphnia dubia	SR	Nom	>90	48	25	Mortality	<24 h	0.683 ± 0.072	Wheelock et al. (2004)
Chaoborus crystallinus	S	Nom	>85	24	15	Mortality	Larvae	0.2 (0.03–0.4)	Stephenson (1982)
Chironomus thummi	S	Nom	>85	24	15	Immobility	Larvae	0.2 (0.1–0.3)	Stephenson (1982)
Cloeon dipterum	S	Nom	>85	24	15	Mortality	Larvae	0.6 (0.3–1)	Stephenson (1982)
Corixa punctata	S	Nom	>85	24	15	Immobility	Adults	0.7 (0.4–2)	Stephenson (1982)
Daphnia magna	SR	Meas	92.3	48	20	Mortality	<24-h old	0.134 (0.114–0.157)	Ward and Boeri (1991)
D. magna	FT	Nom	95.7	48	20	Mortality	<24-h old	0.1615 (0.1344–0.1917)	Wheat and Evans (1994)
Geometric mean								0.147	
Gammarus pulex	S	Nom	>85	24	15	Mortality	3–8 mm	0.1 (0.08–0.2)	Stephenson (1982)

Gyrinus natator	S	Nom	>85	24	15	Immobility	Adults	0.07 (0.04–0.2)	Stephenson (1982)
Hyalella azteca	SR	Meas	>98	96	23	Mortality	7–14 days	0.0021 (0.0017–0.0025)	Weston and Jackson (2009)
H. azteca	SR	Meas	>98	96	23	Mortality	7–14 days	0.0023 (0.0013–0.0035)	Weston and Jackson (2009)
H. azteca	SR	Meas	>98	96	23	Mortality	7–14 days	0.0031 (0.0020–0.0044)	Weston and Jackson (2009)
H. azteca	SR	Nom	97.0	96	23	Mortality	Adults	0.0036 (0.002–0.0049)	Hamer (1997)
Geometric mean								0.0027	
Oncorhynchus mykiss	FT	Meas	91.5	96	12	Mortality	83-day-old juvenile	0.90 (0.72–1.35)	Vaishnav and Yurk (1990)
Oreochromis niloticus	FT	Meas	98.4	96	25	Mortality	0.6–3.0 g	2	Stephenson et al. (1984)
Piona carnea	S	Nom	>85	24	15	Mortality	Adults	0.05 (0.03–0.08)	Stephenson (1982)

All studies were rated relevant and reliable (RR)

S static, *SR* static renewal, *FT* flow through, *95% CI* 95% confidence interval

Table 7 Final chronic toxicity data set for cypermethrin

Species	Test type	Meas/ Nom	Chemical grade (%)	Duration (days)	Temp (°C)	End point	Age/size	NOEC (µg/L)	LOEC (µg/L)	MATC (µg/L)	Reference
Pimephales promelas	FT	Meas	93.1	60	25	Mortality	<48 h	0.077	0.15	0.11	Tapp et al. (1988)

All studies were rated relevant and reliable (RR)

S static, *SR* static renewal, *FT* flow through, *NR* not reported

Table 8 Final acute toxicity data set for λ-cyhalothrin

Species	Test type	Meas/ Nom	Chemical grade (%)	Duration (h)	Temp (°C)	End point	Age/size	LC_{50}/EC_{50} (μg/L) (95% CI)	Reference
Asellus aquaticus	S	Nom	88.0	48	20	Immobility	NR	0.026 (0.018–0.036)	Hamer et al. (1998)
Brachydanio rerio	FT	Meas	88.7	96	25	Mortality	0.70 g, 36 mm	0.64 (0.48–0.90)	Kent and Shillabeer (1997c)
Ceriodaphnia dubia	S	Nom	97.0	48	25	Mortality	<24	0.200 ± 0.090	Wheelock et al. (2004)
Chaoborus sp.	S	Nom	88.0	48	20	Maintenance of body shape/ equilibrium	Larvae	0.0028 (0.0018–0.0041)	Hamer et al. (1998)
Cloeon dipterum	S	Nom	88.0	48	20	Immobility	Nymph	0.038 (0.023–0.093)	Hamer et al. (1998)
Corixa sp.	S	Nom	88.0	48	20	Immobility	NR	0.030 (0.021–0.042)	Hamer et al. (1998)
Cyclops sp.	S	Nom	88.0	48	20	Immobility	NR	0.300 (0.200–0.460)	Hamer et al. (1998)
Daphnia magna	FT	Meas	94.3	72	20	Mortality	<24 h	0.013 (0.010–0.017)	Farrelly and Hamer (1989)
Gammarus pulex	FT	Meas	99.2	96	15	Immobility	5 mm, >3 weeks	0.0059	Hamer et al. (1985a)
Gasterosteus aculeatus	FT	Meas	87.7	96	12	Mortality	0.41 g, 34 mm	0.40 (0.33–0.50)	Long and Shillabeer (1997a)
Hyalella azteca	S	Nom	88.0	48	20	Immobility	NR	0.0023 (0.0010–0.0078)	Hamer et al. (1998)
Hydracarina (Class)	S	Nom	88.0	48	20	Immobility	NR	0.047 (0.033–0.062)	Hamer et al. (1998)
Ictalurus punctatus	FT	Meas	87.7	96	17	Mortality	1.57 g, 48 mm	0.16 (0.13–0.20)	Long and Shillabeer (1997b)

(continued)

Table 8 (continued)

Species	Test type	Meas/Nom	Chemical grade (%)	Duration (h)	Temp (°C)	End point	Age/size	LC_{50}/EC_{50} (μg/L) (95% CI)	Reference
Lepomis macrochirus Rafinesque	FT	Meas	99.0	96	21.9	Mortality	Juvenile	0.106 (0.0855–0.140)	Marino and Rick (2001)
L. macrochirus	FT	Meas	98.0	96	22	Mortality	1.51 g, 38.2 mm	0.21 (0.18–0.25)	Hill (1984b)
Geometric mean								0.15	
Leuciscus idus	FT	Meas	88.7	96	12	Mortality	2.15 g, 53 mm	0.078 (0.056–0.11)	Kent and Shillabeer (1997a)
Oncorhynchus mykiss	FT	Meas	99.0	96	12	Mortality	39 mm, 0.52 g	0.19 (0.16–0.20)	Machado (2001)
O. mykiss	FT	Meas	81.5	96	12	Mortality	43 mm, 1.12 g	0.44 (0.38–0.51)	Tapp et al. (1989)
O. mykiss	FT	Meas	98.0	96	12	Mortality	38.3 mm, 0.83 g	0.24 (0.08–0.70)	Hill (1984a)
Geometric mean								0.27	
Ostracoda (class)	S	Nom	88.0	48	20	Immobility	NR	3.300 (2.100–6.600)	Hamer et al. (1998)
Pimephales promelas	FT	Meas	97.0	96	25	Mortality	Larvae	0.360 (0.252–0.765)	Tapp et al. (1990)
P. promelas	FT	Meas	88.7	96	25	Mortality	0.37 g, 28 mm	0.70 (0.38–1.3)	Kent and Shillabeer (1997d)
Geometric mean								0.50	
Poecilia reticulata	FT	Meas	88.7	96	25	Mortality	0.62 g, 33 mm	2.3 (1.8–3.1)	Kent and Shillabeer (1997b)
Procambarus clarkii	SR	Nom	99.1	96	21.7	Mortality	3-month old	0.16 (0.06–0.27)	Barbee and Stout (2009)

All studies were rated relevant and reliable (RR)

S static, *SR* static renewal, *FT* flow through

Table 9 Final chronic toxicity data set for λ-cyhalothrin

Species	Test type	Meas/ Nom	Chemical grade (%)	Duration (days)	Temp (°C)	End point	Age/size	NOEC (μg/L)	LOEC (μg/L)	MATC (μg/L)	Reference
Daphnia magna	FT	Meas	94.3	21	20	Reproduction (young/female/day)	<24 h	0.00198	0.00350	0.00263	Farrelly and Hamer (1989)
D. magna	SR	Meas	94.3	21	20	Reproduction (young/female/day)	<24 h	0.00375	0.00490	0.00429	Hamer et al. (1985b)
Geometric mean										0.00336	
Pimephales promelas	FT	Meas	97.0	56	25	F_1 survival	F_1 larvae	0.031	0.062	0.044	Tapp et al. (1990)

All studies were rated relevant and reliable (RR)
SR static renewal, *FT* flow through

Table 10 Final acute toxicity data set for permethrin

Species	Test type	Meas/ Nom	Chemical grade (%)	Duration (h)	Temp (°C)	End point	Age/size	LC/EC$_{50}$ (μg/L) (95% CI)	Reference
Ceriodaphnia dubia	S	Nom	99.0	48	25	Mortality	<24 h	0.250 (±119)	Wheelock et al. (2004)
C. dubia	S	Nom	99.3	96	21	Mortality	<24 h	0.652 (0.484–0.856)	Yang et al. (2007)
C. dubia	S	Nom	99.3	96	21	Mortality	<24 h	0.788 (0.545–1.040)	Yang et al. (2007)
C. dubia	S	Nom	99.3	96	21	Mortality	<24 h	0.622 (0.427–0.824)	Yang et al. (2007)
C. dubia	S	Nom	99.3	96	21	Mortality	<24 h	0.772 (0.574–1.013)	Yang et al. (2007)
C. dubia	S	Nom	99.3	96	21	Mortality	<24 h	0.745 (0.568–0.957)	Yang et al. (2007)
C. dubia	S	Nom	99.3	96	21	Mortality	<24 h	0.858 (0.591–1.138)	Yang et al. (2007)
C. dubia	S	Nom	99.3	96	21	Mortality	<24 h	0.571 (0.427–0.740)	Yang et al. (2007)
C. dubia	S	Nom	99.3	96	21	Mortality	<24 h	0.580 (0.407–0.718)	Yang et al. (2007)
C. dubia	S	Nom	99.3	96	21	Mortality	<24 h	0.609 (0.486–0.747)	Yang et al. (2007)
C. dubia	S	Nom	99.3	96	21	Mortality	<24 h	0.570 (0.459–0.689)	Yang et al. (2007)
C. dubia	S	Nom	99.3	96	21	Mortality	<24 h	0.827 (0.669–1.012)	Yang et al. (2007)
C. dubia	S	Nom	99.3	96	21	Mortality	<24 h	0.585 (0.677–0.793)	Yang et al. (2007)
C. dubia	S	Nom	99.3	96	21	Mortality	<24 h	0.849 (0.655–1.085)	Yang et al. (2007)
C. dubia	S	Nom	99.3	96	21	Mortality	<24 h	0.889 (0.666–1.120)	Yang et al. (2007)
C. dubia	S	Nom	99.3	96	21	Mortality	<24 h	0.865 (0.672–1.098)	Yang et al. (2007)
Geometric mean						Mortality		0.664	
Chironomus dilutus	S	Meas	>96	96	23	Mortality	Fourth instar	0.189 (0.131–0.295)	Harwood et al. (2009)
Danio rerio	SR	Nom	90.0	96	23	Mortality	3.0 cm, 0.3 g	2.5 (1.7–3.2)	Zhang et al. (2010)
Daphnia magna	S	Nom	Technical	48	22	Immobility	<24 h	0.32 (0.24–0.44)	LeBlanc (1976)
Erimonax monachus	S	Nom	95.2	96	17	Mortality	NR	1.7	Dwyer et al. (2005)
Etheostoma fonticola	S	Nom	95.2	96	22	Mortality	62 mg, 20.2 mm	3.34 (2.75–4.16)	Dwyer et al. (1999, 2005)
Etheostoma lepidum	S	Nom	95.2	96	22	Mortality	NR	2.71 (2.36–3.13)	Dwyer et al. (1999, 2005)
Hyalella azteca	S	Nom	100.0	96	23	Mortality	Third instar	0.0211	Anderson et al. (2006)

Ictalurus punctatus	S	Nom	92.4	96	21	Mortality	1.2 g, 35 mm	5.4 (3.9–7.4)	Buccafusco (1976a)
Notropis mekistocholas	S	Nom	95.2	96	17	Mortality	NR	4.16	Dwyer et al. (2005)
Oncorhynchus apache	S	Nom	95.2	96	12	Mortality	0.615 g	1.71 (1.3–2.2)	Dwyer et al. (1995, 2005), Sappington et al. (2001)
Oncorhynchus clarki henshawi	S	Nom	95.2	96	12	Mortality	0.46 g	1.58 (1.1–2.2)	Dwyer et al. (1995, 2005), Sappington et al. (2001)
Oncorhynchus mykiss	FT	Meas	91.9	96	15.6	Mortality	Juvenile	7.0	Holcombe et al. (1982)
Orconectes immunis	S	Nom	92.0	96	16.5	Mortality	Juvenile 2 g	0.21 (0.17–0.25)	Paul and Simonin (2006)
Pimephales promelas	S	Nom	95.2	96	22	Mortality	0.41 g	9.38 (6.7–16)	Dwyer et al. (1995, 2005), Sappington et al. (2001)
Procambarus blandingi	FT	Nom	89.1	96	22	Mortality	24 g, 48 mm	0.21 (0.13–0.33)	Buccafusco (1977)
Procloeon sp.	S	Nom	100.0	48	23	Mortality	0.5–1 cm	0.0896	Anderson et al. (2006)
Salmo salar	S	Nom	92.4	96	12	Mortality	1 g, 35 mm	1.5 (1.1–2.0)	Buccafusco (1976b)
Xyrauchen texanus	S	Nom	95.2	96	22	Mortality	0.32 g	5.95 (4.6–7.7)	Dwyer et al. (1995, 2005), Sappington et al. (2001)

All studies were rated relevant and reliable (RR)
S static, *SR* static renewal, *FT* flow through

Table 11 Final chronic toxicity data set for permethrin

Species	Test type	Meas/ Nom	Chemical grade (%)	Duration (days)	Temp (°C)	End point	Age/size	NOEC (μg/L)	LOEC (μg/L)	MATC (μg/L)	Reference
Brachycentrus americanus	FT	Meas	Technical	21	15	Mortality	Larvae	–	–	LC_{50}: 0.17 (0.09–0.34)	Anderson (1982)
Daphnia magna	FT	Meas	98.6	21	20	Reproduction	<24 h	0.039	0.084	0.057	Kent et al. (1995a)
D. magna	FT	Meas	98.6	21	20	Length	<24 h	0.039	0.084	0.057	Kent et al. (1995a)
D. magna										0.057	
Pimephales promelas	FT	Meas	92.0	32	25	Mortality	4–5-day-old larvae	0.66	1.4	0.96	Spehar et al. (1983)

All studies were rated relevant and reliable (RR)
S static, *SR* static renewal, *FT* flow through, *NR* not reported

distribution was selected as the best fit for both λ-cyhalothrin and permethrin based on maximum likelihood estimation using the BurrliOZ software (CSIRO 2001). Fit tests based on cross validation and Fisher's combined test found no significant lack of fit for bifenthrin, cyfluthrin, λ-cyhalothrin, or permethrin, with $X^2_{2n} > 0.199$ for these four compounds (calculations shown in the Supporting Material http://extras.springer.com/). The Burr III distribution was initially selected as the best fit for the cypermethrin data set, but this distribution did not provide a satisfactory fit based on the fit test ($\chi^2_{2n} = 0.000014$; calculations shown in the Supporting Material http://extras.springer.com/); so a log-logistic distribution, which is less likely to overfit the data, was fit to the cypermethrin data set instead.

Acute values were derived from the distributions, including fifth percentiles (median and lower 95% confidence limit), as well as first percentiles (median and lower 95% confidence limit). The median fifth percentile is the most robust of the four distributional estimates, and is therefore the estimate recommended for criteria calculation.

Bifenthrin Log-Logistic Distribution

HC5 fitting parameters: $\alpha = -0.661$; β (median) $= 0.4872$, β (lower 95% CI) $= 0.9328$

Fifth percentile, 50% confidence limit: 0.00803 μg/L
Fifth percentile, 95% confidence limit: 0.000391 μg/L
First percentile, 50% confidence limit: 0.00126 μg/L
First percentile, 95% confidence limit: 0.0000113 μg/L
Recommended acute value: 0.00803 μg/L (median fifth percentile)

$$\text{Acute criterion} = \frac{\text{Acute value}}{2}. \tag{1}$$

Bifenthrin acute criterion = 0.004 μg/L

Cyfluthrin Log-Logistic Distribution

HC5 fitting parameters: $\alpha = -0.7446$; β (median) $= 0.5478$; β (lower 95% CI) $= 1.04898$

Fifth percentile, 50% confidence limit: 0.00439 μg/L
Fifth percentile, 95% confidence limit: 0.000147 μg/L
First percentile, 50% confidence limit: 0.000547 μg/L
First percentile, 95% confidence limit: 0.0000027 μg/L
Recommended acute value: 0.00439 μg/L (median fifth percentile)
Cyfluthrin acute criterion = 0.002 μg/L

Cypermethrin Log-Logistic Distribution

HC_5 fitting parameters: $\alpha = -0.6601$, β(median) $= 0.4199$, β(lower 95% CI) $= 0.6768$

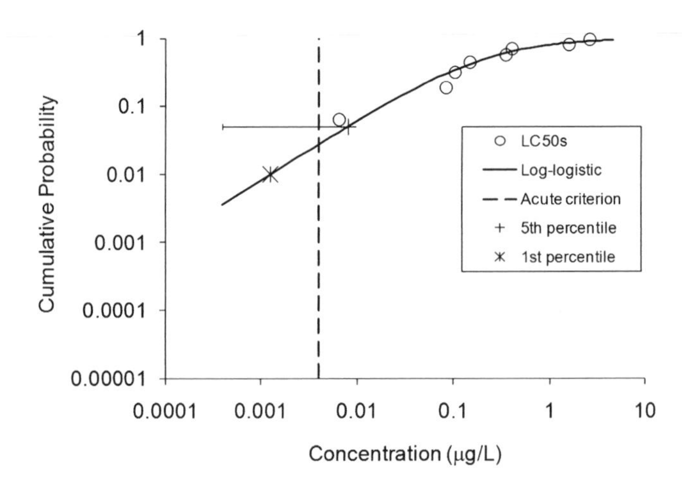

Fig. 1 Plot of bifenthrin species mean acute values and fit of the log-logistic distribution. The graph shows the median fifth and first percentiles with the lower 95% confidence limit on the fifth percentile and the acute criterion at 0.004 μg/L

Fifth percentile, 50% confidence limit: 0.0127 μg/L
Fifth percentile, 95% confidence limit: 0.00222 μg/L
First percentile, 50% confidence limit: 0.00257 μg/L
First percentile, 95% confidence limit: 0.000170 μg/L
Recommended acute value: 0.0127 μg/L (median fifth percentile)
Cypermethrin acute criterion = 0.006 μg/L

λ-Cyhalothrin Burr III Distribution

Fit parameters: $b = 0.232356$; $c = 1.100750$; $k = 0.596085$ (likelihood = −4.987264)

Fifth percentile, 50% confidence limit: 0.00243 μg/L
Fifth percentile, 95% confidence limit: 0.000501 μg/L
First percentile, 50% confidence limit: 0.000208 μg/L
Recommended acute value: 0.002432 μg/L (median fifth percentile)
λ-Cyhalothrin acute criterion = 0.001 μg/L

Permethrin Burr III Distribution

Fit parameters: $b = 7.80465$; $c = 6.599725$; $k = 0.07608$ (likelihood = 35.742158)

Fifth percentile, 50% confidence limit: 0.020008 μg/L
First percentile, 50% confidence limit: 0.000811 μg/L
Recommended acute value: 0.020008 μg/L (median fifth percentile)
Permethrin acute criterion = 0.01 μg/L

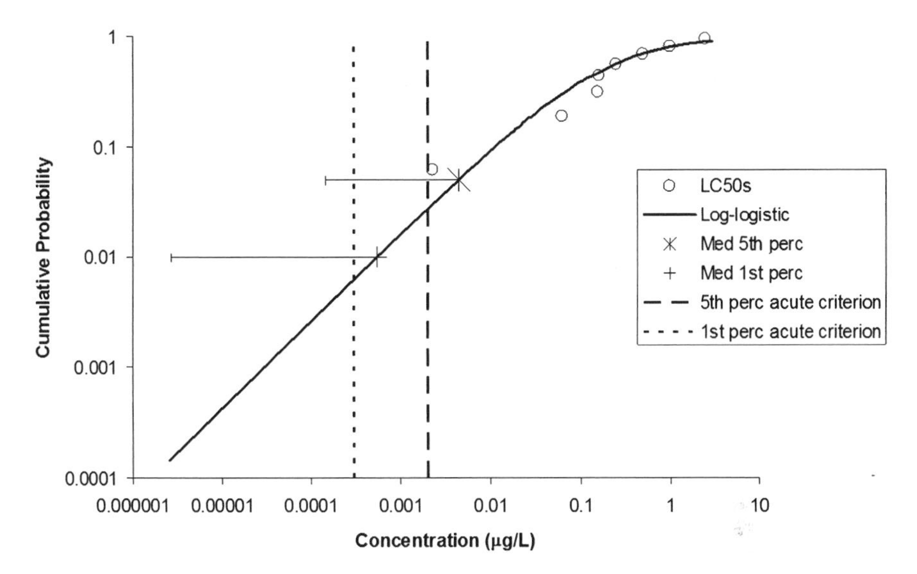

Fig. 2 Plot of cyfluthrin species mean acute values and fit of the log-logistic distribution. The graph shows the median fifth and first percentile values with the lower 95% confidence limits and the acute criteria calculated using both the median fifth percentile value and the median first percentile value

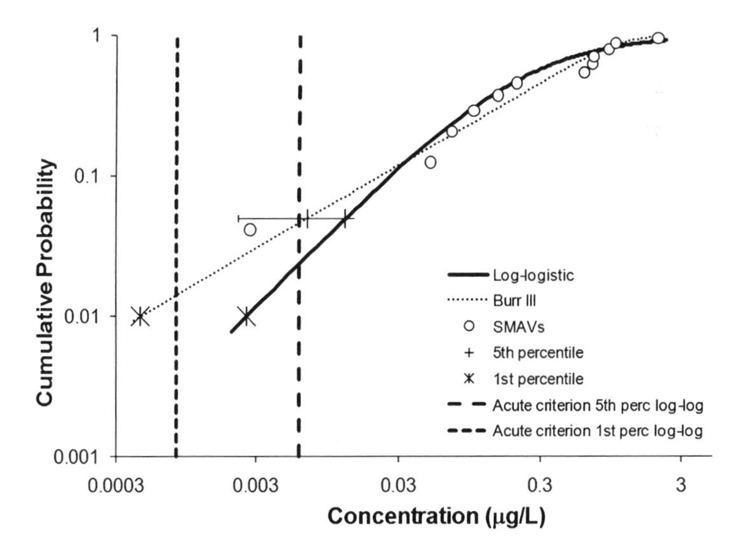

Fig. 3 Plot of species mean acute values for cypermethrin and fit of the log-logistic and Burr Type III distributions. The graph shows the median fifth and first percentiles for both distributions with the lower 95% confidence limit for the median fifth percentile.on the log-logistic, and the acute criteria calculated using both the median fifth and first percentiles of the log-logistic distribution

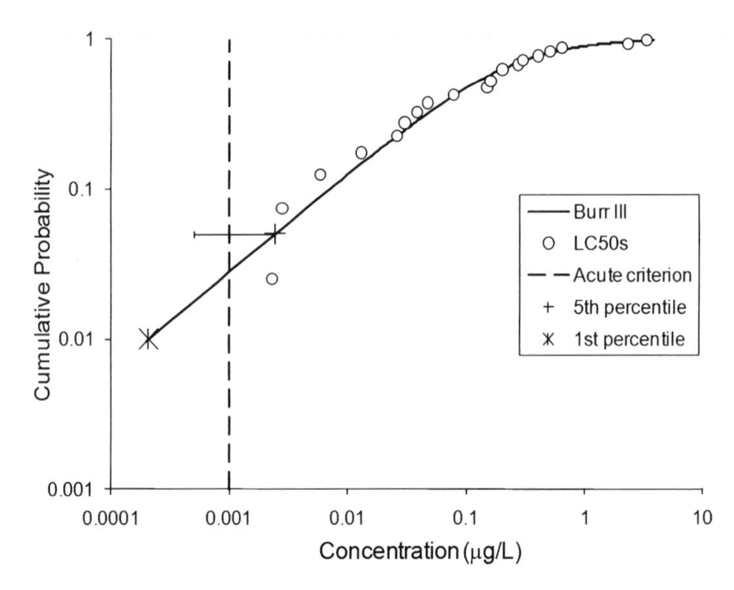

Fig. 4 Plot of species mean acute values for λ-cyhalothrin and fit of the Burr III distribution. The graph shows the median fifth and first percentile values with the lower 95% confidence limit of the fifth percentile and the acute criterion at 0.001 μg/L

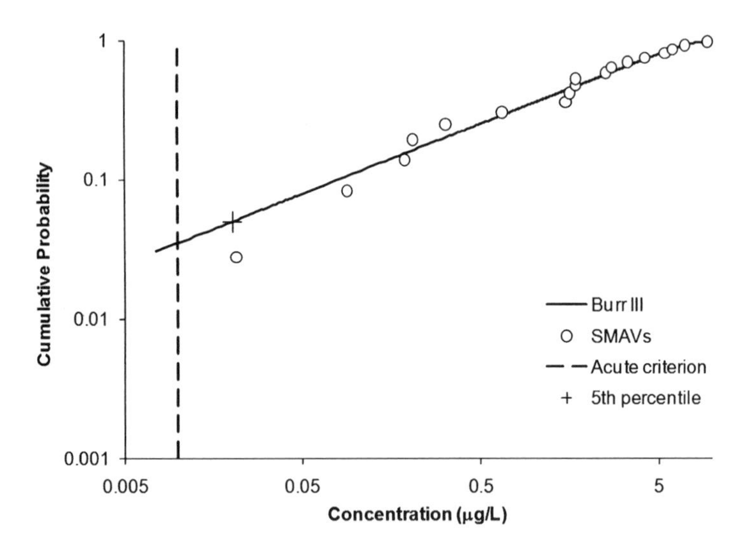

Fig. 5 Plot of species mean acute values for permethrin and fit of the Burr III distribution. The graph shows the median fifth percentile and the acute criterion at 0.01 μg/L

The fits of the distributions to the acute data sets are shown in Figs. 1–5, in cumulative probability plots. Because there is variability between the first significant digits of the median and lower 95% confidence limit estimates, the final criteria are reported with one significant figure. Although a lower 95% confidence limit could not be calculated for the permethrin distribution for comparison, the permethrin acute criterion is also reported with one significant digit because there was variability in the first digit of the fifth percentiles generated in the fit test. Later in this chapter, the acute criteria for cyfluthrin and cypermethrin are recalculated using lower percentile estimates because the criteria for these compounds calculated with the median fifth percentile acute values were not protective when compared to data for sensitive species, threatened and endangered species, and multispecies ecosystem-level studies.

5 Chronic Criterion Calculations

The chronic data sets of all five pyrethroids were limited and did not include data that met the five taxa requirements of the SSD procedure. Instead, the ACR procedure (TenBrook et al. 2010) was used for chronic criteria calculation, which is based on the ACR procedure in the USEPA (1985) method, but also includes a default ACR, when ACR data is also limited. For bifenthrin and cypermethrin, only one or two of the five SSD taxa requirements were satisfied (Tables 3 and 7), and none of these values could be paired with an appropriate corresponding acute toxicity value to calculate ACRs. There were also no appropriate saltwater data to use for ACR calculation; thus, these chronic criteria were calculated with the default ACR of 12.4 (TenBrook et al. 2010). Three of the five cyfluthrin taxa requirements were satisfied: a species in the family Salmonidae (*Oncorhynchus mykiss*), a warm water fish (*Pimephales promelas*), and a planktonic crustacean (*Daphnia magna*). Each of these three chronic values were paired with appropriate corresponding acute toxicity values, which satisfied the three family requirements (a fish, an invertebrate, and another sensitive species) for the ACR procedure with measured toxicity data. There were two λ-cyhalothrin chronic toxicity values, satisfying two of the five taxa requirements: a warm water fish (*P. promelas*) and a planktonic crustacean (*D. magna*). Both freshwater chronic values were paired with corresponding acute toxicity values to calculate ACRs, and paired data for the saltwater species *Cyprinodon variegatus* was used to complete the third family requirement for the ACR procedure. While two chronic values were available for permethrin, neither could be paired with appropriate acute data, but paired data for the saltwater species *Americamysis bahia* was available. This ACR was combined with two default ACRs to complete the three ACR requirements, allowing for the calculation of a final ACR for permethrin.

To calculate species mean ACRs (SMACRs) for each species, the acute LC_{50} was divided by the chronic maximum acceptable toxicant concentration (MATC)

Table 12 Calculation of the final acute-to-chronic ratio for cyfluthrin

Species	LC_{50} (μg/L)	Acute reference	Chronic end point	MATC (μg/L)	Chronic reference	ACR (LC_{50}/MATC)
Daphnia magna	0.160	Burgess (1990)	Reproduction/ length	0.02864	Forbis et al. (1984)	5.58659
Oncorhynchus mykiss	0.2512	Bowers (1994), Gagliano and Bowers (1994)	Biomass/ weight	0.0133	Carlisle (1985)	18.88970
Pimephales promelas	2.49	Rhodes et al. (1990)	Various	0.20149	Rhodes et al. (1990)	12.35793
Multispecies ACR = geomean (individual ACRs)						10.27

Table 13 Calculation of the species mean acute-to-chronic ratios for λ-cyhalothrin

Species	LC_{50} (μg/L)	Chronic end point	MATC (μg/L)	Reference	ACR (LC_{50}/MATC)
Cyprinodon variegatus	0.81	Weight	0.31	Hill et al. (1985)	2.6129
Daphnia magna	0.013	Reproduction (young/ female/day)	0.00263	Farrelly and Hamer (1989)	4.9430
Pimephales promelas	0.36	FI survival	0.044	Tapp et al. (1990)	8.1818
Multispecies ACR = geomean (individual ACRs)					4.73

Table 14 Acute-to-chronic ratios used for derivation of the permethrin chronic criterion

Species	LC_{50} (μg/L)	Acute reference	Chronic end point	MATC (μg/L)	Chronic reference	SMACR (LC_{50}/MATC)
Americamysis bahia	0.075	Thompson (1986)	Mortality	0.016	Thompson et al. (1989)	4.6875
Default						12.4[a]
Default						12.4[a]
Multispecies ACR = geomean (individual ACRs)						8.96592

[a] The derivation and source data of the default ACR are described in the UCDM (TenBrook et al. 2010)

for a given species. The final ACRs for cyfluthrin (10.27) and λ-cyhalothrin (4.73) were calculated as the geometric mean of all of the SMACRs in each ACR data set and the final ACR for permethrin (8.96592) was calculated with one SMACR and two default ACRs (Tables 12–14) while the default ACR of 12.4 was used for bifenthrin and cypermethrin. Chronic criteria were calculated by dividing the recommended acute value (median fifth percentile) by the final ACR. Later in this chapter, the cyfluthrin and cypermethrin are adjusted downward to be protective based on comparisons to data for sensitive, threatened, and endangered species and ecosystem-level studies.

Bifenthrin chronic criterion calculated with the acute median fifth percentile estimate:
Fifth percentile, 50% confidence limit: 0.00803 μg/L

$$
\begin{aligned}
\text{Chronic criterion} &= \frac{\text{Recommended acute value}}{\text{ACR}}, \qquad (2)\\
&= \frac{0.0243\ \mu\text{g/L}}{12.4},\\
&= 0.0006\ \ \mu\text{g/L},\\
&= 0.6\ \text{ng/L}.
\end{aligned}
$$

Cyfluthrin chronic criterion calculated with the acute median fifth percentile estimate:
Fifth percentile, 50% confidence limit: 0.00439 μg/L

$$
\begin{aligned}
\text{Chronic criterion} &= \frac{0.00439\ \mu\text{g/L}}{10.27},\\
&= 0.0004\ \ \mu\text{g/L},\\
&= 0.4\ \text{ng/L}.
\end{aligned}
$$

Cypermethrin chronic criterion calculated with the acute median fifth percentile estimate:
Fifth percentile, 50% confidence limit: 0.0126904 μg/L

$$
\begin{aligned}
\text{Chronic criterion} &= \frac{0.0126904\ \mu\text{g/L}}{12.4},\\
&= 0.001\ \ \mu\text{g/L},\\
&= 1\ \text{ng/L}.
\end{aligned}
$$

λ-Cyhalothrin chronic criterion calculated with the acute median fifth percentile estimate:
Fifth percentile, 50% confidence limit: 0.00243 μg/L

$$
\begin{aligned}
\text{Chronic criterion} &= \frac{0.00243\ \mu\text{g/L}}{4.73},\\
&= 0.0005\ \ \mu\text{g/L},\\
&= 0.5\ \text{ng/L}.
\end{aligned}
$$

Permethrin chronic criterion calculated with the acute median fifth percentile estimate:
Fifth percentile, 50% confidence limit: 0.020008 μg/L

$$
\begin{aligned}
\text{Chronic criterion} &= \frac{0.020008\ \mu\text{g/L}}{8.96592},\\
&= 0.002\ \ \mu\text{g/L},\\
&= 2\ \text{ng/L}.
\end{aligned}
$$

6 Bioavailability

Although pyrethroids are not very soluble in water, aquatic organisms are very sensitive to the pyrethroids and toxicity does occur. Pyrethroids have been found as the cause of toxicity in surface waters in the California Central Valley (Phillips et al. 2007; Weston et al. 2009a; Weston and Lydy 2010). This toxicity is believed to occur primarily from the fraction of the compound that is dissolved in the water, not from the compound that is associated with particulate phases. For example, Surprenant (1988) demonstrated that bifenthrin from spiked soil samples was available at concentrations sufficient to cause toxicity to *D. magna* that were housed in a separate container from the sediment, but shared the same recirculating water (however, dissolved particles could have been involved in this exposure).

Numerous studies demonstrate that the uptake and toxicity of pyrethroids are greatly reduced when solids or dissolved organic matter (DOM) are present (Day 1991; DeLorenzo et al. 2006; Lajmanovich et al. 2003; Muir et al. 1985, 1994; Smith and Lizotte 2007; Yang et al. 2006a, b, c, 2007). These studies indicate that bound pyrethroids are unavailable, and thus nontoxic to aquatic organisms. It has been presumed that the pyrethroids primarily sorb to the organic carbon phase of solids or DOM, and Hunter et al. (2008) demonstrated that sediment OC-normalized concentrations of permethrin were highly correlated with the uptake of permethrin by *Chironomus dilutus* (formerly *C. tentans*), which supports this assumption. Yet Yang et al. (2007) did not find a direct correlation between dissolved organic carbon (DOC) content and uptake or toxicity of pyrethroids, indicating that partitioning is not solely dependent on the quantity of DOC, but is also dependent on the quality of the DOC. Consequently, to accurately estimate pyrethroid sorption to DOC and particulate OC in whole water, site-specific partition coefficients would be preferred.

Alternately, the bioavailable fraction of pyrethroids can be estimated by measuring only the freely dissolved concentration using solid-phase microextraction (SPME). Yang et al. (2006a, 2007) reported that organism uptake was closely mimicked by SPME results and that the aqueous concentration of pyrethroids measured by SPME correlated well with the variations in uptake and toxicity with various DOM. Xu et al. (2007) clearly demonstrated that it is the freely dissolved aqueous concentration of pyrethroid that is bioavailable when they tested bifenthrin and cyfluthrin toxicity to *C. tentans* in 10-day sediment exposures with three types of sediment. The researchers reported LC_{50}s for five phases: bulk sediment, OC-normalized sediment, bulk pore water, DOC-normalized pore water, and freely dissolved pyrethroid. The LC_{50}s calculated for each of the five phases varied greatly, and varied between sediments for all phases tested except the freely dissolved, indicating that toxicity of the freely dissolved phase is independent of site-specific characteristics. The LC_{50}s based on the freely dissolved concentrations were at least an order of magnitude lower than those based on bulk pore water concentrations that included DOC, indicating that toxicity may be greatly underestimated if bioavailability is not taken into account. Based on these myriad studies, it can be concluded that the freely dissolved concentration is the most accurate predictor of toxicity and that bound pyrethroids were unavailable to the studied organisms.

However, bound pyrethroids can continue to desorb into the water column for long periods of time because pyrethroids have long equilibration times (~30 days; Bondarenko et al. 2006) and environmental systems are usually not at true equilibrium. The fraction of chemical that is potentially available to an organism is known as the bioaccessible fraction, and it has been linked to biological effects (Semple et al. 2004; You et al. 2011). Benthic organisms, such as *Hyalella azteca*, may be at greater risk because of their exposure to pore water and close proximity to sediments, where dissolved concentrations may persist.

Additionally, the role of dietary exposure on bioavailability of pyrethroids has not been considered. In the tests performed by Yang et al. (2006a, b) with *Ceriodaphnia dubia* and *D. magna*, organisms were not fed during the test. Organisms living in contaminated waters may also be ingesting food with sorbed hydrophobic compounds that can be desorbed by digestive juices (Mayer et al. 2001). The effects of dietary exposure may also be species specific, depending on typical food sources; some species may have greater interaction with particles, increasing their exposure. Palmquist et al. (2008) examined the effects due to dietary exposure of the pyrethroid esfenvalerate on three aqueous insects with different feeding functions: a grazing scraper (*Cinygmula reticulata* McDunnough), an omnivore filter feeder (*Brachycentrus americanus* Banks), and a predator (*Hesperoperla pacifica* Banks). The researchers observed adverse effects in *C. reticulata* and *B. americanus* after feeding on esfenvalerate-laced food sources and that none of the three insects avoided the contaminated food. The effects included reduced growth and egg production of *C. reticulata* and abandonment and mortality in *B. americanus*. Stratton and Corke (1981) tested toxicity of permethrin to *D. magna* with and without feeding of algae, and found that mortality at 24 h was significantly increased when daphnids were fed, although mortality at 48 h was not affected. The authors proposed that permethrin may have been ingested by the daphnids if it was sorbed on the algal cells, and caused increased toxicity, although the same effect was not seen when bacteria were provided as a food source. These limited studies indicate that ingestion may be an exposure route, but it is not currently possible to incorporate this exposure route into criteria compliance assessment.

The studies above suggest that the freely dissolved fraction of pyrethroids is the primary bioavailable fraction and that this concentration is the best indicator of toxicity; thus, it is recommended that the freely dissolved fraction is directly measured or calculated based on site-specific information for compliance assessment. The most direct way to determine compliance would be to measure the pyrethroid concentration in the dissolved phase to determine the total bioavailable concentration. SPME has shown to be a good predictor of pyrethroid toxicity in many studies (Bondarenko et al. 2007; Bondarenko and Gan 2009; Hunter et al. 2008; Xu et al. 2007; Yang et al. 2006a, b, c, 2007). Bondarenko and Gan (2009) reported method detection limits of 1.0 ng/L for bifenthrin, 2.0 ng/L for cyfluthrin, 2.0 ng/L for cypermethrin, 2.4 ng/L for λ-cyhalothrin, 2.0 ng/L for cis-permethrin, and 3.0 for trans-permethrin, and Li et al. (2009) reported method detection limits of 0.2 ng/L for bifenthrin, 0.2 for cyhalothrin, 0.9 ng/L for cyfluthrin, 1.0 ng/L for cypermethrin, and 1.2 ng/L for permethrin using SPME. Analytical detection limits may create a problem for criteria compliance because most of these reported detection limits are above the derived criteria, meaning it is possible that one of these pyrethroids could be present in toxic amounts, yet below the detection limit so that an excursion is not identified. Filtration of suspended solids is not recommended for determining criteria compliance because pyrethroids have been demonstrated to adsorb to glass fiber filters by Gomez-Gutierrez et al. (2007). They found that on average 58% of a 50 ng/L solution of permethrin was lost on the filter; this magnitude of loss may be critical for determining compliance at environmental concentrations.

If the freely dissolved concentration is not directly measured, the following equation can be used to translate total pyrethroid concentrations measured in whole water to the associated dissolved pyrethroid concentrations:

$$C_{\text{dissolved}} = \frac{C_{\text{total}}}{1 + ((K_{\text{OC}} \times [\text{SS}])/f_{\text{oc}}) + (K_{\text{DOC}} \times [\text{DOC}])}, \quad (3)$$

where $C_{\text{dissolved}}$ is the concentration of chemical in dissolved phase (μg/L), C_{total} is the total concentration of chemical in water (μg/L), K_{OC} is the OC–water partition coefficient (L/kg), [SS] is the concentration of suspended solids in water (kg/L), f_{oc} is the fraction of OC in suspended sediment in water, [DOC] is the concentration of dissolved organic carbon in water (kg/L), and K_{DOC} is the OC–water partition coefficient (L/kg) for DOC.

To determine compliance by this calculation, site-specific data are necessary, including K_{OC}, K_{DOC}, concentration of suspended solids, concentration of DOC, and fraction of OC in the suspended solids. If all of these site-specific data, including the partition coefficients, are not available, then this equation should not be used for compliance determination. Site-specific data are required because the sorption of pyrethroid to suspended solids and DOM depends on the physical and chemical properties of the suspended solids resulting in a range of K_{OC} and K_{DOC} values, as discussed earlier in this section.

The freely dissolved pyrethroid concentration is recommended for determination of criteria compliance because the literature suggests that the freely dissolved concentrations are the most accurate predictor of toxicity. Environmental managers may choose an appropriate method for determining the concentration of freely dissolved pyrethroid. If environmental managers choose to measure whole water concentrations for criteria compliance assessment, the bioavailable fraction will likely be overestimated.

7 Chemical Mixtures

Pyrethroids often co-occur in the environment (Trimble et al. 2009; Werner and Moran 2008), and various other chemical mixtures are ubiquitous in surface waters. Because the presence of other chemicals can add to or alter the toxicity of another given chemical, it is important to examine the effects of chemical mixtures on individual pyrethroid toxicity. Although chemical interactions are rarely straightforward, the concentration addition model is recommended for chemicals with the same toxicological mode of action. All pyrethroids have a similar mode of action in that they bind to and prolong the opening of voltage-dependent ion channels, causing convulsions, paralysis, and death (Brander et al. 2009). The three studies that tested toxicity of pyrethroid mixtures found that the effects were generally well-predicted by the concentration addition model (Barata et al. 2006; Brander et al. 2009; Trimble et al. 2009). Overall, the concentration addition model should be used by following either the toxic unit or relative potency factor approach to determine criteria compliance when multiple pyrethroids are present.

Barata et al. (2006) observed slight antagonism for *D. magna* survival for λ-cyhalothrin—deltamethrin mixtures, but the deviation from additivity was attributed to a few unexpected extreme values for joint survival effects, as most observed effects were within a factor of 2 of the effects predicted by the concentration addition model. Brander et al. (2009) tested mixture toxicity of cyfluthrin and permethrin, and found slight antagonism for the binary mixture, but additivity was demonstrated when piperonyl butoxide (PBO) was added. Brander et al. (2009) offered several explanations for the observed antagonism between the two pyrethroids. Permethrin is a type I pyrethroid and cyfluthrin is a type II pyrethroid, and type II pyrethroids may be able to outcompete type I pyrethroids for binding sites, which is known as competitive agonism; or binding sites may be saturated so that complete additivity is not observed. They also note that cyfluthrin is metabolized more slowly than permethrin, so cyfluthrin can bind longer. PBO may remove this effect because the rate of metabolism of both pyrethroids is reduced in its presence. To examine if pyrethroid mixture toxicity is additive with a more comprehensive study design, Trimble et al. (2009) performed sediment toxicity tests with *H. azteca* in three binary combinations: type I–type I (permethrin–bifenthrin), type II–type II (cypermethrin–λ-cyhalothrin), and type I–type II (bifenthrin–cypermethrin). The toxicity of these combinations was predicted with the concentration addition model, with model deviations within a factor of 2, indicating that in general pyrethroid mixture toxicity is additive.

PBO is commonly added to pyrethroid insecticide treatments because it is known to increase the toxic effects of pyrethroids (Weston et al. 2006). Many studies have demonstrated that the addition of PBO at a concentration that would be nonlethal on its own increases the toxicity of pyrethroids (Brander et al. 2009; Brausch and Smith 2009; Hardstone et al. 2007, 2008; Kasai et al. 1998; Paul and Simonin 2006; Paul et al. 2005, 2006; Rodriguez et al. 2005; Singh and Agarwal 1986; Xu et al. 2005). Several of these studies report single-species interaction coefficients (*K*; also called synergistic ratios) for pyrethroids and PBO ranging from 1.35 (*D. magna*; Brausch and Smith 2009) to 60 (snails; Singh and Agarwal 1986). While many studies report interaction coefficients for synergism of PBO, none of them reported interaction coefficients for multiple PBO concentrations; so a relationship between PBO concentration and *K* cannot be determined for any given species. In addition, no multispecies interaction coefficients are available; thus, there is no accurate way to account for synergism with PBO in compliance determination.

Mixture effects with pyrethroids and various other chemicals have also been studied and are summarized here, but there are currently no multispecies interaction coefficients available for these combinations. Binary mixtures of λ-cyhalothrin with deltamethrin and cadmium demonstrated additivity (Barata et al. 2006, 2007). Mixtures with various fungicides have been investigated and some synergism has been demonstrated. Norgaard and Cedergreen (2010) reported synergism with equitoxic mixtures of the fungicides and α-cypermethrin, yielding interaction coefficients ranging from 1.4 to 27, while other ratios tested resulted in interaction coefficients ranging from 0.41 to 37. Adam et al. (2009) also reported synergism for mixtures of fungicides and cypermethrin, which are often found in combination in wood preservatives. Permethrin in combination with propoxur, a carbamate, demonstrated synergism, which the authors propose is due to the complementary modes of action acting on different parts of the nervous system (Corbel et al. 2003). The thiocarbamate pesticide cartap appears to be antagonistic when combined with cypermethrin as no toxicity was observed in tests with *D. magna* and *Oryzias latipes*, when the concentrations of each chemical tested in combination were higher than the reported EC/LC_{50} values for the single chemicals (Kim et al. 2008). Gartenstein et al. (2006) reported synergism for cypermethrin in binary combinations with diflubenzuron and diazinon, but the combination of all three compounds produced an antagonistic effect. Zhang et al. (2010) tested mixtures of permethrin with the organophosphates dichlorvos or phoxim and reported that the toxicity of binary combinations was additive.

No studies on aquatic organisms were found in the literature that could provide a quantitative means to consider mixtures of pyrethroids with other classes of pesticides. Although there are examples of nonadditive toxicity, multispecies interaction coefficients are not available for any pyrethroid, and therefore the concentrations of nonadditive chemicals cannot be used for criteria compliance.

8 Water Quality Effects

Temperature has been reported to be inversely proportional to the aquatic toxicity and bioavailability of pyrethroids (Miller and Salgado 1985; Werner and Moran 2008). In fact, the increase of toxicity of pyrethroids with decreasing temperature has been used to implicate pyrethroids as the source of toxicity in environmental samples (Phillips et al. 2004; Weston et al. 2009b). The inverse relationship between temperature and pyrethroid toxicity is likely due to the increased sensitivity of an organism's sodium channels at lower temperatures (Narahashi et al. 1998).

Enhanced toxicity of cyfluthrin to larval fathead minnows (*P. promelas*) at lower temperatures was demonstrated by Heath et al. (1994). Sublethal cyfluthrin concentrations reduced the ability of fish to tolerate temperatures both higher and lower than standard conditions. The toxicities of six aqueous pyrethroids were 1.33- to 3.63-fold greater at 20°C compared to 30°C for mosquito larvae (Cutkomp and Subramanyam 1986). Harwood et al. (2009) tested permethrin toxicity to *C. dilutus* in an aqueous exposure at 13 and 23°C, and reported a

3.2-fold decrease of the 96-h LC_{50} at the lower temperature. Kumaraguru and Beamish (1981) reported that for small trout the toxicity of permethrin increased by a factor of 10 with a decrease in temperature from 20 to 5°C, but showed little change from 10 to 5°C. These studies indicate that the enhanced toxic effects of pyrethroids at lower temperature may not be as accurately represented by the results of typical laboratory toxicity tests, which tend to be run at warmer temperatures, 20–23°C (USEPA 1996a, b, 2000) than those of the habitats of coldwater fishes, about 15°C or lower (Sullivan et al. 2000).

The toxicity of sediments contaminated with pyrethroids was more than twice as toxic when tested at 18°C compared to 23°C (Weston et al. 2008). Weston et al. (2008) used a toxicity identification evaluation (TIE) procedure to determine the effect of temperature reduction (18 vs. 23°C) on toxicity of a particular environmental sediment sample to *H. azteca*. These results are not directly applicable for use in water quality criteria compliance because they were sediment exposures and used environmental samples, instead of an exposure to a pure compound.

Unfortunately, there are limited data in which aquatic exposures with relevant species were used, making it unfeasible to quantify the relationship between the toxicity of these five pyrethroids and temperature for water quality criteria at this time. Information regarding the effects of pH or other water quality parameters on pyrethroid toxicity was not identified, but based on the physical–chemical properties of these compounds they are not expected to be affected by these parameters.

9 Sensitive Species

Data for particularly sensitive species found in the acceptable (RR) and supplemental (RL, LR, LL) data sets (Tables S8–S12, Supporting Material http://extras.springer.com/) were compared to the criteria. There are some species represented in the supplemental data set that are not represented in the acceptable data set, and it is possible that data at the extreme sensitive end of the data set could be below the criteria derived using the median fifth percentiles. The bifenthrin acute criterion of 4 ng/L is below the lowest freshwater SMAV in the bifenthrin data sets (6.5 ng/L for *H. azteca*), and the chronic criterion of 0.6 ng/L is below the lowest freshwater SMCV in the data sets (1.9 ng/L for *D. magna*), so these criteria appear to be protective based on the available data. For λ-cyhalothrin, the acute and chronic criteria calculated with the acute median fifth percentile (1 and 0.5 ng/L, respectively) are both below all of the freshwater toxicity values in the respective acute and chronic data sets. The lowest LC_{50} is 2.3 ng/L for *H. azteca* while the lowest freshwater MATC is 2.63 ng/L for *D. magna*. For bifenthrin and λ-cyhalothrin, there are toxicity values equal to or below the derived criteria for the saltwater species *A. bahia*, but the criteria were not adjusted because they are only intended to protect freshwater species. The permethrin acute criterion (10 ng/L) is below the lowest acute value in the acute data sets (21.1 ng/L for *H. azteca*; Anderson et al. 2006). The permethrin chronic criterion (2 ng/L) is below all of the chronic values in the available data sets (16 ng/L for *A. bahia*; Thompson et al. 1989).

The lowest SMAV in the cyfluthrin RR data set (Table 4) was 2.3 ng/L for *H. azteca*, which is approximately equal to the derived acute criterion of 2 ng/L. Based on the available data, the criterion derived using the median fifth percentile acute value is not protective of *H. azteca*; therefore, the next lowest acute value was used to calculate the cyfluthrin criteria. The acute and chronic cyfluthrin criteria calculations using the median first percentile acute value are as follows:

Recommended acute value: 0.000547 μg/L (median first percentile)

$$\begin{aligned}\text{Cyfluthrin acute criterion} &= \frac{0.000547\ \mu\text{g/L}}{2},\\ &= 0.0003\ \mu\text{g/L (0.3 ng/L)}.\\ \text{Cyfluthrin chronic criterion} &= \frac{0.000547\ \mu\text{g/L}}{10.27},\\ &= 0.00005\ \mu\text{g/L (0.05 ng/L)}.\end{aligned}$$

The cyfluthrin chronic criterion calculated with the median first percentile (0.05 ng/L) is below the lowest MATC in the data sets of 0.27 ng/L for *A. bahia*.

The derived cypermethrin acute criterion (0.006 μg/L) is higher than one SMAV in the RR acute data set, 0.0027 μg/L for *H. azteca* (Table 6). The *H. azteca* SMAV is the geometric mean of four values, three from a study in which concentrations were measured (Weston and Jackson 2009), all of which are lower than the acute criterion of 0.006 μg/L. Thus, the next lowest estimate from the log-logistic distribution (median first percentile) was used to derive the cypermethrin acute and chronic criteria as follows:

Recommended acute value: 0.0025723 μg/L (median first percentile)

$$\begin{aligned}\text{Cypermethrin acute criterion} &= \frac{0.0025723\ \mu\text{g/L}}{2},\\ &= 0.001\ \mu\text{g/L (1 ng/L)}.\\ \text{Cypermethrin chronic criterion} &= \frac{0.0025723\ \mu\text{g/L}}{12.4},\\ &= 0.0002\ \mu\text{g/L (0.2 ng/L)}.\end{aligned}$$

There is one supplemental datum (96-h $EC_{50} = 0.6$ ng/L for *D. magna*) that is below the adjusted cypermethrin acute criterion, but this toxicity value was not based on measured concentrations, and this species is represented in the RR data set with an SMAV that indicates that it is protected by the acute criterion. There are two supplemental MATCs that are below the adjusted chronic criterion of 0.2 ng/L (MATCs of 0.00063 and 0.063 ng/L for *D. magna*; Kim et al. 2008), but they are based on nominal concentrations, and it is recommended that criteria should only be adjusted based on toxicity values calculated with measured concentrations.

10 Ecosystem-Level Studies

Toxicity data from multispecies studies that more closely mimic ecosystems can yield different results than single-species effect studies, so community-level study data were compared to the criteria derived from single-species studies to ensure that the criteria are protective of ecosystems. A total of 28 studies addressing effects on microcosms, mesocosms, and model ecosystems were rated acceptable (R or L reliability rating) for the five selected pyrethroids (ratings listed in Table S13, Supporting Material http://extras.springer.com/). None of the bifenthrin, cyfluthrin, or permethrin studies reported ecosystem-level NOECs or no-effect concentrations (NECs) to which the chronic criteria could be directly compared.

Data in three of the bifenthrin studies (Drenner et al. 1993; Hoagland et al. 1993; Surprenant 1988) included toxic effects at concentrations ranging from 20 to 3,150 ng/L, which are well above the derived chronic criterion (0.6 ng/L). Sherman (1989) reported toxic effects for several invertebrates and fish in a pond receiving runoff contaminated with bifenthrin, but the effects did not correlate well with aqueous bifenthrin concentrations. Average pond concentrations fluctuated from slightly above 1 to 10 ng/L, but could not be linked to the occurrence of toxicity.

Authors of all the cyfluthrin studies reported toxic effects at applied or measured concentrations that were far above the chronic criterion (Gunther and Herrmann 1986; Johnson 1992; Johnson et al. 1994; Kennedy et al. 1990; Morris 1991; Morris et al. 1994). Toxic effects were observed in all of the studies, especially on aquatic macroinvertebrates, but it is not possible to assess if effects would have occurred if lower concentrations were tested, closer to the chronic criterion of 0.05 ng/L.

Several studies consisted of single concentrations of cypermethrin (0.01–24,000 μg/L) that were well above the chronic criterion in pond or marine mesocosms, followed by measurement of the recovery of the invertebrate communities. Toxic effects were observed particularly for insects and crustaceans, and some populations did not recover during the posttreatment observation periods (Crossland 1982; Farmer et al. 1995; Maund et al. 2009; Medina et al. 2004). The study by Maund et al. (2009) simulated natural reinvasion in some microcosms by adding invertebrates to the enclosures post treatment; in these microcosms, there was a general recovery of invertebrate populations in approximately 100 days. In contrast, the microcosms that received no additional organisms showed only limited recovery after 16 weeks of observation. These results indicate that small, isolated, or heavily impacted waterbodies will likely recover more slowly than waterbodies that are only partially impacted or are near other unimpacted waterbodies from which organisms can immigrate.

Friberg-Jensen et al. (2003) calculated cypermethrin NECs for crustaceans, copepods, and cladocerans ranging from 0.02 to 0.07 μg/L in enclosures set in a lake. These NECs are all significantly higher than the chronic criterion of 0.0004 μg/L. They also reported that rotifers, protozoans, bacteria, periphyton plankton, and periphytic algae all proliferated after treatment with cypermethrin, in response to the decreased populations of grazers. A sister paper, describing effects for the same experiment, reported an NEC of 0.01 μg/L for copepod nauplii (Wendt-Rasch et al. 2003). This paper also reported significant changes to species composition of the aforementioned communities at nominal concentrations greater than 0.13 μg/L.

Several λ-cyhalothrin studies reported community NOECs to which the calculated criteria may be compared. Van Wijngaarden et al. (2006) and Roessink et al. (2005) reported various community-level NOECs that were season- and trophic-system-dependent, the lowest being $<$10 μg/L, and Schroer et al. (2004) reported a community-level NOEC of 10 ng/L. Schroer et al. (2004) also calculated a community-level criterion of 4.1 ng/L while the criterion calculated based on laboratory single-species data was 2.7 ng/L. The UCDM chronic criterion (0.5 ng/L) is below the reported NOECs for this set of studies by at least a factor of 20.

Hill et al. (1994) investigated the effects of λ-cyhalothrin on artificial pond mesocosms containing microbes, algae, macrophytes, zooplankton, macroinvertebrates, and fish. λ-cyhalothrin was applied at three rates as a spray and as a soil–water slurry to simulate runoff. Few effects were observed for most taxa, but macroinvertebrates and zooplankton were adversely affected at the highest rate; macroinvertebrates experienced some effects at the middle rate as well. Measured aqueous concentrations of λ-cyhalothrin ranged from 3 to 98 ng/L, in the mesocosms treated at the highest rate, and 2 to 10 ng/L in those treated at the middle rate. λ-cyhalothrin was not detected in the ponds treated at the lowest rate. The method detection limit reported in this study ranges from 2 to 3 ng/L, so it is possible that λ-cyhalothrin was present at lower concentrations when reported as nondetects. This study indicates that the derived chronic criterion of 0.5 ng/L should be protective of macroinvertebrates and zooplankton because it is likely similar to the actual concentrations in the ponds treated at the lowest rate.

Several study authors reported significant macroinvertebrate mortality and drift due to exposure to λ-cyhalothrin (Farmer et al. 1995; Lauridsen and Friberg 2005; Rasmussen et al. 2008; Wendt-Rasch et al. 2004), particularly for *Gammarus* species. Farmer et al. (1995) sprayed pond mesocosms with λ-cyhalothrin (measured at 2 ng/L, 1 h post treatment for the lower rate) and reported that *Gammarus* spp. abundance was significantly reduced compared to controls. Rasmussen et al. (2008) demonstrated that *Gammarus pulex* exposed to 10.65 ng/L λ-cyhalothrin (nominal) for 90 min and then transferred to clean water drifted significantly more than controls ($p<0.0001$). Phytoplankton and algae productivity increased in response to λ-cyhalothrin exposure (Farmer et al. 1995; Rasmussen et al. 2008; Wendt-Rasch et al. 2004) likely due to the decrease in macroinvertebrate populations, as macroinvertebrates are known to graze on algae. Lauridsen and Friberg (2005) examined macroinvertebrate drift in outdoor experimental channels with two insect species and *G. pulex*. Catastrophic drift was observed for all three species during the 1-h pulse exposure and 2–3-h post exposure. Drift of *G. pulex* was significantly affected at 1 ng/L (nominal), and it should be noted that the measured concentrations may have been even lower. While several studies indicate that *Gammarus* species experience lethal and sublethal effects due to λ-cyhalothrin exposures at concentrations near the chronic criterion, none of them reported toxicity values (e.g., NOEC, EC_x) or measured concentrations at or below the derived chronic criterion; thus, the chronic criterion is not adjusted downward at this time.

In all permethrin studies, adverse effects were reported on aquatic organisms, but they all used formulations and test concentrations (0.02–100 μg/L) that were significantly higher than the chronic criterion of 0.002 μg/L. In two studies, increased drifting in model riverine systems was reported after exposure to permethrin for some invertebrate species (Poirier and Surgeoner 1988; Werner and Hilgert 1992), and another model riverine study reported that snails and water thyme (*Elodea*) were both adversely affected at permethrin concentrations of 4 and 20 μg/L (Lutnicka et al. 1999). Several pond exposures also demonstrated adverse effects on various aquatic invertebrates, including some populations that did not recover during the posttreatment observation period (Conrad et al. 1999; Coulon 1982; Yasuno et al. 1988). Conrad et al. (1999) dosed small artificial ponds with permethrin (1–100 μg/L nominal Picket® formulation) and conducted bioassays with chironomids, which were compared to laboratory sediment toxicity tests with *Chironomus riparius*. The chironomid responses of reduced larval density and adult emergence were not predicted by bulk sediment chemistry, sediment toxicity tests, or laboratory bioassay results—all three measurements underestimated the acute effects. Toxicity to *C. riparius* in the field was best predicted by acute water-only toxicity test data, indicating that the primary exposure route is via the water column. This study supports measurement of the truly dissolved fraction for criteria compliance and indicates the relevance of water quality criteria for protection of aquatic life.

11 Threatened and Endangered Species

Data for species listed as threatened or endangered were examined to ensure that the criteria are protective of these species. Both the US Fish and Wildlife Service federal list of threatened and endangered plant and animal species (USFWS 2010) and the California state list of threatened and endangered plant and animal species (the California Department of Fish and Game (CDFG) 2010a, 2010b) were consulted for this evaluation.

There are ten evolutionarily significant units of *O. mykiss* listed as federally threatened or endangered, and this species is represented in the data sets of all five pyrethroids that were examined. There are SMAVs for this species in all five acute data sets ranging from 0.119 to 7 μg/L, which are well above the derived acute criteria for these compounds. The λ-cyhalothrin acute data set also includes *Gasterosteus aculeatus*, of which a subspecies (*G.a. williamsoni*) is endangered. The acute permethrin data set includes seven additional listed species: *Oncorhynchus clarki henshawi*, *Etheostoma fonticola*, *Erimonax monachus*, *Notropis mekistocholas*, *Oncorhynchus apache*, *Salmo salar*, and *Xyrauchen texanus*. All of these acute toxicity values were used in criteria calculation and are well above the derived criteria; hence, there is no evidence that the criteria are underprotective of these species. The only chronic toxicity value for a listed species was an MATC for *O. mykiss* in the cyfluthrin data set of 0.0133 μg/L, which is much higher than the chronic criterion of 0.00005 μg/L, indicating that the chronic criterion is protective of this species.

All of the acute data sets include species that are not listed but are in the same family or genus as some of those that are. These species were used as surrogates to estimate toxicity values for related TES with the USEPA interspecies correlation estimation software (Web-ICE v. 3.1; Raimondo et al. 2010). Unfortunately, the available bifenthrin and cyfluthrin SMAVs were below the model minimum input values, so toxicity values could not be predicted for bifenthrin or cyfluthrin. *O. mykiss* was used to predict λ-cyahlothrin, cypermethrin, and permethrin acute toxicity values for up to 13 species in the Salmonidae family (Tables S14–S16, Supporting Material http://extras.springer.com/). The predicted acute toxicity values ranged from 0.262 to 0.576 μg/L for cyfluthrin, 0.860 to 1.31 μg/L for cypermethrin, and 3.48 to 11.88 μg/L for permethrin, which are all more than one order of magnitude above their respective acute criteria.

One caveat of this evaluation is that the only TES in the measured or predicted data sets for these pyrethroids were fish, which are relatively insensitive compared to aquatic amphipods and insects. There were no data for TES in these more sensitive taxa, so it is not clear if the derived criteria are protective of these species. No single-species plant studies were found in the literature for use in criteria derivation for any of these pyrethroids, so no estimation could be made for plants on the state or federal endangered, threatened, or rare species lists. Phytoplanktons were unaffected by bifenthrin in a pond study (Sherman 1989); however, bifenthrin seemed to be beneficial in some instances and harmful in others, as reported in a mesocosm study that monitored primary productivity, green algae, chlorophyll, and other end points for photosynthetic organisms (Hoagland et al. 1993). Based on the mode of action, plants should be relatively insensitive to pyrethroids and the calculated criteria should be protective of aquatic plants.

12 Bioaccumulation

Chemicals in surface waters can accumulate in organisms from both the water and food items, which is called bioaccumulation, and eventually the chemicals can move up the food chain from prey to predator. Potential bioaccumulation was assessed to ensure that the derived criteria are set at concentrations that are not likely to cause toxicity due to bioaccumulation. Bifenthrin, cyfluthrin, cypermethrin, λ-cyhalothrin, and permethrin have similar physical–chemical characteristics (Table 1), including molecular weight $<1{,}000$ and log-normalized octanol–water partition coefficients ($\log K_{ow}$) >3.0 L/kg, which indicate that all five compounds have the potential to bioaccumulate.

Low-to-moderate bioaccumulation of pyrethroids has been documented in the literature. For example, wild-caught brown trout (*Salmo trutta*), captured in a

British stream, was found to have accumulated an average 25.4 μg/kg of cyfluthrin and as high as 109 μg/kg in tissues, even though no cyfluthrin could be detected in the water column (Bonwick et al. 1996). Additionally, Surprenant (1986) reported that elimination of bifenthrin from bluegill tissues was very slow, i.e., after 42 days of depuration, fish tissue concentrations of bifenthrin were reduced by about half.

Because the pyrethroids have the potential to bioaccumulate, available data were used to estimate aqueous concentrations not expected to lead to harmful bioaccumulation. Analogous calculations were not done for human consumption of aquatic organisms because there are no tolerance or USFDA action levels for fish tissue (USFDA 2000) for any of these compounds. To calculate an aqueous NOEC, the dietary NOEC of an oral predator (mallard duck; studies listed in Table S17, Supporting Material http://extras.springer.com/) is divided by the bioaccumulation factor (BAF) for a fish. If a BAF is not available for a fish, it can be calculated as the product of the bioconcentration factor (BCF) and a biomagnification factor (BMF) such that BAF $=$ BCF $\times$ BMF. BCFs are a measure of the uptake of a chemical by an organism from water alone while BMFs are a measure of the uptake of a chemical by an organism from food sources. BCFs for the pyrethroids of interest varied widely among different species, and were dependent on what portion of an organism was analyzed, with BCFs ranging from 2.6 to 3,280,000 (Table S18, Supporting Material http://extras.springer.com/).

For bifenthrin, one dietary NOEC was available for reproductive effects on mallard duck of 75 mg/kg (Roberts et al. 1986). No BAFs or BMFs were identified for fish, so the BCF of 28,000 L/kg bifenthrin for whole *P. promelas* (McAllister 1988) and a default BMF of 10, based on the log K_{ow} (TenBrook et al. 2010), were used to estimate a BAF as follows:

$$\mathrm{NOEC_{water}} = \frac{\mathrm{NOEC_{oral\_predator}}}{\mathrm{BCF_{food\_item}} \times \mathrm{BMF_{food\_item}}}. \tag{4}$$

The resulting $\mathrm{NOEC_{water}}$ for bifenthrin is 267 ng/L, which is well above the chronic criterion of 0.6 ng/L, indicating that bifenthrin at concentrations equal to or below the chronic criterion will not likely cause harm via bioaccumulation.

This calculation was also performed for the other four pyrethroids. For cyfluthrin, the highest BCF of 854 L/kg for *Lepomis macrochirus* (Carlisle and Roney 1984), a default BMF of 10, and the lowest dietary NOEC for a mallard of 250 mg/kg (Carlisle 1984) were used for a conservative estimation. The $\mathrm{NOEC_{water}}$ estimated for cyfluthrin using this data was 29 μg/L, which is above the aqueous solubility of cyfluthrin (2.3 μg/L; Laskowski 2002). For cypermethrin, the values used in Eq. 4 were the highest fish BCF of 821 L/kg for *O. mykiss*, a default BMF of 10, and a dietary toxicity value for mallard duck of 50 mg/kg, although this dietary NOEC was reported as greater than (>) 50 mg/kg (USEPA 2008). These values resulted in an $\mathrm{NOEC_{water}}$ for mallard of 6.09 μg/L, which is above the aqueous solubility of cypermethrin (4 μg/L; Laskowski 2002). An $\mathrm{NOEC_{water}}$ of 1.34 μg/L was calculated for λ-cyhalothrin with the highest reported BCF of 2,240 L/kg for whole fish *Cyprinus carpio* (Yamauchi et al. 1984), a default BMF of 10, and an oral predator dietary NOEC of 30 mg/kg for mallard duck (Beavers et al. 1990). This $NOEC_{water}$ is significantly larger than the λ-cyhalothrin chronic criterion of 0.0005 μg/L. Finally, this calculation was completed for permethrin using the highest fish BCF of 2,800 L/kg for *P. promelas*, a default BMF of 10, and the dietary NOEC for mallard duck of 125 mg/kg, giving an $NOEC_{water}$ of 4.46 μg/L, which is nearing the aqueous solubility of 5.7 μg/L. Based on these conservative calculations, these pyrethroids are not likely to cause adverse effects on terrestrial wildlife due to bioaccumulation if their concentrations do not exceed the derived chronic criteria.

13 Assumptions, Limitations, and Uncertainties

Data limitations and important assumptions are reviewed here so that environmental decision makers have information about the accuracy and confidence in the criteria. Assumptions and limitations inherent in the methodology are summarized in the UCDM (TenBrook et al. 2010). The principal limitation for these five pyrethroids was a dearth of chronic data, particularly for the most sensitive species, amphipods and other invertebrates. There were no appropriate paired acute and chronic data for bifenthrin or cypermethrin to calculate ACRs, so the default ACR was used, while measured ACRs were available for cyfluthrin, λ-cyhalothrin, and permethrin. The acute criterion for cyfluthrin calculated with the median fifth percentile was almost identical to the lowest SMAV in the RR data set while the acute criterion for cypermethrin calculated with the median fifth percentile was higher than the lowest SMAV in the RR data set, so these criteria were adjusted downward to be more protective using less robust acute values. There are inherent assumptions in the use of an SSD (TenBrook et al. 2010), and the various distributional estimates can be used to assess uncertainty in the derived criteria for each compound. Of the data that were available for these compounds, not all were from flow-through tests that reported measured concentrations, which can cause overestimation of toxicity values, because pyrethroids are highly sorptive.

Other conspicuous data gaps were regarding temperature effects and mixture toxicity, especially with PBO; additional data on these topics should lead to quantifiable correlations, and these considerations should be added to criteria compliance when available. Also, pyrethroids are known to partition to sediments, and if federal or state sediment quality standards become available for these compounds, partitioning should be predicted based on the derived water quality criteria to ensure that these aqueous concentrations are not leading to potentially harmful sediment concentrations.

14 Comparison to Existing Criteria

To date, the USEPA has not calculated water quality criteria for bifenthrin, cyfluthrin, cypermethrin, λ-cyhalothrin, or permethrin. The CDFG composed a risk assessment report for several pyrethroids, including bifenthrin, cypermethrin, and permethrin (Siepmann and Holm 2000). CDFG concluded that there were insufficient data to calculate criteria for bifenthrin using the USEPA (1985) method, and instead they reported the lowest acute and chronic toxicity values for guidance. The lowest genus mean acute value (GMAV) for bifenthrin was 3.97 ng/L for *A. bahia*, which is only slightly below the UCDM acute criterion of 4 ng/L; it can be noted that *A. bahia* is a saltwater species, which may be more sensitive than freshwater species. The lowest bifenthrin MATC in the CDFG report was 60 ng/L for *P. promelas*, which would not be protective of *D. magna* with an MATC of 1.9 ng/L (Table 3). The CDFG risk assessment reported interim acute criteria of 2 ng/L for cypermethrin and 30 ng/L for permethrin, which are both higher than the acute criteria calculated using the UCDM by factors of 2 and 3, respectively. Chronic criteria were not calculated for cypermethrin or permethrin because there was insufficient data.

The Netherlands has done generic risk assessment for several pyrethroids and maximum permissible concentrations (MPCs) have been calculated using the Dutch criteria derivation methodology. An MPC is defined as the concentration in the environment above which the risk of adverse effects is considered unacceptable to ecosystems and they are harmonized across media (Crommentuijn et al. 2000). MPCs are analogous to chronic water quality criteria and are used as the basis for setting environmental quality standards in the Netherlands. The Dutch MPCs for bifenthrin, cypermethrin, and permethrin are 1.1, 0.09, and 0.2 ng/L, respectively (Crommentuijn et al. 2000). These values were calculated with a modified EPA method in which an assessment factor ranging from 10 to 1,000 is applied to the lowest available toxicity value. The bifenthrin MPC of 1.1 ng/L is larger than the chronic criterion derived via the UCDM of 0.6 ng/L by a factor of 1.8, but there are no data to indicate that the MPC would be underprotective. The cypermethrin and permethrin MPCs are smaller than the UCDM criteria by a factor of 2.2 and 10, respectively.

In Canada, an interim freshwater quality guideline was derived for permethrin by applying a safety factor of 0.1 to the most sensitive LOEC, which was for *Pteronarcys dorsata* (Anderson 1982). The interim aquatic life guideline for permethrin was derived as 4 ng/L, which is larger than the UCDM chronic criterion of 2 ng/L, but there are no data to indicate that 4 ng/L would be underprotective (CCME 2006). Quebec has also derived its own interim acute criterion for permethrin of 44 ng/L and an interim chronic criterion of 13 ng/L (Guay et al. 2000); the interim acute criterion is larger than the lowest SMAV in the UCD data set and would not be protective of *H. azteca*. In the UK, short-term and long-term predicted NECs (PNECs), analogous to acute and chronic criteria, were recently derived for permethrin using assessment factors (Lepper et al. 2007). The short-term PNEC of 10 ng/L was derived by applying an assessment factor of 10 to the LC_{50} for the mayfly *Hexagenia bilineata*, and the long-term PNEC of 1.5 ng/L was derived by applying an assessment factor of 20 to an LOEC for the caddisfly *B. americanus* (Lepper et al. 2007). The currently adopted long-term (chronic) environmental quality standard (EQS) in the UK for permethrin is 10 ng/L. There are also proposed long-term and short-term EQSs for cypermethrin of 0.1 and 0.4 ng/L, respectively, which are lower than the existing EQSs of 0.2 and 2.0 ng/L, respectively (UKTAG 2008).

15 Comparison to the USEPA (1985) Method

More pyrethroid toxicity data are available now than when CDFG derived criteria for bifenthrin, cypermethrin, and permethrin (Siepmann and Holm 2000) using the USEPA (1985) method. To compare the UCDM criteria to those generated using the USEPA (1985) method, the data sets gathered for this article were used to generate example USEPA criteria for these compounds. The five acute taxa requirements of the SSD procedure in the UCDM were fulfilled for each of these five pyrethroids. There are three additional taxa requirements in the USEPA acute method, as follows:

1. A third family in the phylum Chordata (e.g., fish, amphibian)
2. A family in a phylum other than Arthropoda or Chordata (e.g., Rotifera, Annelida, Mollusca)
3. A family in any order of insect or any phylum not already represented

These three additional requirements were not met for any of these compounds. The bifenthrin, λ-cyhalothrin, and permethrin data sets do not contain any species in a phylum other than Arthropoda or Chordata, but met all of the other taxa requirements. The CDFG has calculated criteria for compounds with incomplete data sets if the missing taxa requirements are known to be relatively insensitive to the compound of interest. The only data available for organisms not in the phyla Arthropoda or Chordata were for saltwater mollusks (*Crassostrea virginica* and *Crassostrea gigas*), which were very insensitive to bifenthrin and λ-cyhalothrin—EC_{50}s could not be calculated for these species because of solubility limits or no responses were observed at the highest concentrations tested (Thompson 1985; Ward 1986a, 1986b, 1987)—so example criteria were calculated for bifenthrin, λ-cyhalothrin, and permethrin. The cyfluthrin and cypermethrin acute data sets were missing two of the additional requirements, so example criteria were not calculated for these compounds.

Acute criteria were calculated by fitting the log-triangular distribution to the acute bifenthrin, λ-cyhalothrin, and permethrin data sets (Tables 2, 8, and 10) and are reported with two significant figures, according to the USEPA (1985) method. The USEPA (1985) method fits the SSD to *genus* mean acute values while the UCDM uses *species* mean acute values, so the UCDM data sets were altered when necessary to calculate genus mean acute values.

Example acute criterion = Final acute value/2

Bifenthrin : Example final acute value (fifth percentile) = 0.0009543 μg/L
Example acute criterion = 0.00048 μg/L

λ-Cyhalothrin: Example final acute value (fifth percentile) = 0.001845 μg/L
Example acute criterion = 0.00092 μg/L

Permethrin : Example final acute value (fifth percentile) = 0.039001 μg/L
Example acute criterion = 0.010 μg/L

The bifenthrin example acute criterion (0.48 ng/L) is almost one order of magnitude lower than the acute criterion calculated by the UCDM (4 ng/L). The λ-cyhalothrin example acute criterion (0.92 ng/L) is almost identical to the acute criterion calculated using the Burr Type III distribution of the UCDM (1 ng/L), and the permethrin example acute criterion (10 ng/L) is identical to the UCDM acute criterion (10 ng/L).

To calculate chronic criteria according to the USEPA (1985) method for compounds with limited chronic data such as the pyrethroids, an ACR procedure is used, which is very similar to the ACR procedure in the UCDM. The ACR procedure cannot be used for cyfluthrin and cypermethrin because acute criteria were not calculated for these compounds. The EPA ACR procedure requires data for three ACRs, which were not available for bifenthrin or permethrin. For λ-cyhalothrin, the same three SMACRs calculated for the UCDM (Table 13) were calculated according to the USEPA (1985) methodology to give a final λ-cyhalothrin ACR of 4.73. The λ-cyhalothrin chronic criterion was calculated by dividing the final acute value by the final ACR:

Example chronic criterion = Final acute value/Final ACR
λ-cyhalothrin example chronic criterion = 0.00039 μg/L

The λ-cyhalothrin example chronic criterion (0.39 ng/L) differs by less than a factor of 2 from the one recommended by the UCDM (0.5 ng/L).

This comparison of criteria calculated using the UCDM and USEPA (1985) method highlights the limitations of the USEPA method. According to the USEPA method, acute criteria could not be calculated for cyfluthrin or cypermethrin, and acute criteria were only calculated for bifenthrin, λ-cyhalothrin, and permethrin by making exceptions for the taxa requirements, and chronic criteria could not be calculated for bifenthrin, cyfluthrin, or cypermethrin. The λ-cyhalothrin acute data set was large and the criteria calculated by the two methods were very similar (1 ng/L vs. 0.92 ng/L). When large data sets are available, criteria calculated using the two methods have been similar, e.g., chlorpyrifos and diazinon (Palumbo et al. (2012)), because the calculation methods in these cases are very similar. When large data sets are not available or data sets are missing a USEPA taxa requirement, the UCDM is able to generate criteria, where the USEPA method gives no results, e.g., malathion (Palumbo et al. (2012)) and cyfluthrin.

16 Final Criteria Statements

The inputs for the final criteria statement are listed in Table 15.

Aquatic life should not be affected unacceptably if the 4-day average concentration of [1] does not exceed [2] μg/L ([3] ng/L) more than once every 3 years on the average and if the 1-h average concentration does not exceed [4] μg/L ([5] ng/L) more than once every 3 years on the average. Mixtures of [1] and other pyrethroids should be considered in an additive manner (see Sect. 7).

Table 15 Final numeric criteria for the five pyrethroids

1 Compound	2 Chronic riterion (μg/L)	3 Chronic criterion (ng/L)	4 Acute criterion (μg/L)	5 Acute criterion (ng/L)
Bifenthrin	0.004	4	0.0006	0.6
Cyfluthrin	0.00005	0.05	0.0003	0.3
Cypermethrin	0.0002	0.2	0.001	1
λ-Cyhalothrin	0.0005	0.5	0.001	1
Permethrin	0.002	2	0.01	10

It is recommended that the freely dissolved pyrethroid concentration is measured for criteria compliance because this appears to be the best predictor of the bioavailable fraction.

17 Summary

Aquatic life water quality criteria were derived for five pyrethroids using a new methodology developed by the University of California, Davis (TenBrook et al. 2010). This methodology was developed to provide an updated, flexible, and robust water quality criteria derivation methodology specifically for pesticides. To derive the acute criteria, log-logistic SSDs were fitted to the medium-sized bifenthrin, cyfluthrin, and cypermethrin acute toxicity data sets while the λ-cyhalothrin and permethrin acute data sets were larger, and Burr Type III SSDs could be fitted to these data sets. A review of the cyfluthrin acute criterion revealed that it was not protective of the most sensitive species in the data set, *H. azteca*, so the acute value was adjusted downward to calculate a more protective criterion. Similarly, the cypermethrin criteria were adjusted downward to be protective of *H. azteca*. Criteria for bifenthrin, λ-cyhalothrin, and permethrin were calculated using the median fifth percentile acute values while the cyfluthrin and cypermethrin criteria were calculated with the next lowest acute value (median first percentile). Chronic data sets were limited in all cases, so ACRs were used for chronic criteria calculations, instead of statistical distributions. Sufficient corresponding acute and chronic data were not available for bifenthrin, cypermethrin, or permethrin, so a default ACR was used to calculate these chronic criteria while measured ACRs were used for cyfluthrin and λ-cyhalothrin. A numeric scoring system was used to sort the acute and chronic data, based on relevance and reliability, and the individual study scores are included in the Supporting Information.

According to the USEPA (1985) method, the data sets gathered for these five pyrethroids would not be sufficient to calculate criteria because they were each missing at least one of the eight taxa required by that method. The USEPA (1985) method generates robust and reliable criteria, and the goal of creating the UCDM was to create a method that also yields statistically robust criteria, but with more flexible calculation methods to accommodate pesticide data sets of varied sizes and diversities. Using the UCDM, acute and chronic water quality criteria were derived for bifenthrin (4 and 0.6 ng/L, respectively), cyfluthrin (0.3 and 0.05 ng/L, respectively), cypermethrin (1 and 0.2 ng/L, respectively), λ-cyhalothrin (1 and 0.5 ng/L, respectively), and permethrin (10 and 2 ng/L, respectively). Water quality criteria for these five pyrethroids can be used by environmental managers to control the increasing problem of surface water contamination by pesticides.

Acknowledgments We thank the following reviewers: D. McClure (CRWQCB-CVR), J. Grover (CRWQCB-CVR), S. McMillan (CDFG), J. P. Knezovich (Lawrence Livermore National Laboratory), X. Deng (CDPR), and E. Gallagher (University of Washington). Funding for this project was provided by the California Regional Water Quality Control Board, Central Valley Region (CRWQCB-CVR). The contents of this document do not necessarily reflect the views and policies of the CRWQCB-CVR, nor does mention of trade names or commercial products constitute endorsement or recommendation for use.

References

Adam O, Badot P-M, Degiorgi F, Crini G (2009) Mixture toxicity assessment of wood preservative pesticides in the freshwater amphipod *Gammarus pulex* (L.). Ecotoxicol Environ Saf 72:441–449.

Aldenberg T (1993) ETX 1.3a. A program to calculate confidence limits for hazardous concentrations based on small samples of toxicity data. National Institute of Public Health and the Environment (RIVM), Bilthoven, The Netherlands.

Amweg EL, Weston DP, You J, Lydy MJ (2006) Pyrethroid insecticides and sediment toxicity in urban creeks from California and Tennessee. Environ Sci Technol 40:1700–1706.

Anderson RL (1982) Toxicity of fenvalerate and permethrin to several nontarget aquatic invertebrates. Environ Entomol 11:1251–1257.

Anderson BS, Phillips BM, Hunt JW, Connor V, Richard N, Tjeerdema RS (2006) Identifying primary stressors impacting macroinvertebrates in the Salinas River (California, USA): Relative effects of pesticides and suspended particles. Environ Poll 141:402–408.

Barata C, Baird DJ, Nogueira AJA, Soares AMVM, Riva MC (2006) Toxicity of binary mixtures of metals and pyrethroid insecticides to *Daphnia magna* Straus. Implications for multi-substance risks assessment. Aquat Toxicol 78:1–14.

Barata C, Baird DJ, Nogueira AJA, Agra AR, Soares AMVM (2007) Life-history responses of *Daphnia magna* Straus to binary mixtures of toxic substances: Pharmacological versus ecotoxicological modes of action. Aquat Toxicol 84:439–449.

Barbee GC, Stout MJ (2009) Comparative acute toxicity of neonicotinoid and pyrethroid insecticides to non-target crayfish (*Procambarus clarkii*) associated with rice-crayfish crop rotations. Pestic Manage Sci 65:1250–1256.

Beavers JB, Hoxter KA, Jaber MJ (1990) PP321: A one-generation reproduction study with the mallard (*Anas platyrhynchos*). Wildlife International Ltd. Project no. 123–143 for ICI Agrochemicals, Easton, MD. Submitted to U.S. Environmental Protection Agency. USEPA MRID: 41512101.

Bondarenko S, Gan J (2009) Simultaneous measurement of free and total concentrations of hydrophobic compounds. Environ Sci Technol 43:3772–3777.

Bondarenko S, Putt A, Kavanaugh S, Poletika N, Gan JY (2006) Time dependence of phase distribution of pyrethroid insecticides in sediment. Environ Toxicol Chem 25:3148–3154.

Bondarenko S, Spurlock F, Gan J (2007) Analysis of pyrethroids in sediment pore water by solid-phase microextraction. Environ Toxicol Chem 26:2587–2593.

Bonwick GA, Yasin M, Hancock P, Baugh PJ, Williams JHH, Smith CJ, Armitage R, Davies DH (1996) Synthetic pyrethroid insecticides in fish: Analysis by gas chromatography–mass spectrometry operated in the negative ion chemical ionization mode and ELISA. Food Agr Immunol 8:185–194.

Bowers LM (1994) Acute toxicity of ^{14}C-cyfluthrin to Rainbow Trout (*Oncorhynchus mykiss*) under flow-through conditions. Miles Incorporated Agriculture Division, Research and Development Dept, Environmental Research Section, Stilwell, KS. Submitted to U.S. Environmental Protection Agency. USEPA MRID: 45426705. CDPR ID: 50317-173.

Brander SM, Werner I, White JW, Deanovic LA (2009) Toxicity of a dissolved pyrethroid mixture to *Hyalella azteca* at environmentally relevant concentrations. Environ Toxicol Chem 28:1493–1499.

Brausch JM, Smith PN (2009) Development of resistance to cyfluthrin and naphthalene among *Daphnia magna*. Ecotoxicology 18:600–609.

Breckenridge CB, Holden L, Sturgess N, Weiner M, Sheets L, Sargent D, Soderlund DM, Choi JS, Symington S, Clark JM, Burr S, Ray D (2009) Evidence for a separate mechanism of toxicity for the Type I and the Type II pyrethroid insecticides. Neurotoxicol 30 S: S17–S31.

Buccafusco RJ (1976a) Acute Toxicity of PP-557 technical to channel catfish (*Ictalurus punctatus*). EG&G Bionomics, Wareham, MA. CDPR ID: study number 15147.

Buccafusco RJ (1976b) Acute toxicity of PP-557 technical to Atlantic salmon (*Salmo salar*). EG&G Bionomics, Wareham, MA. CDPR ID: 00083085, study number 15150.

Buccafusco RJ (1977) Acute toxicity of permethrin technical (PP 557) to crayfish (*Procambarus blandingi*). EG&G Bionomics, Wareham, MA. CDPR ID study number 15140.

Budd R, Bondarenko S, Haver D, Kabashima J, Gan J (2007) Occurrence and bioavailability of pyrethroids in a mixed land use watershed. J Environ Qual 36:1006–1012.

Burgess D (1989) Chronic toxicity of ^{14}C-FMC 54800 to *Daphnia magna* under flow-through test conditions. Analytical Bio-Chemistry Laboratories, Inc. FMC Study No: A88-2649. Columbia, MO. Submitted to U.S. Environmental Protection Agency. EPA MRID: 41156501.

Burgess LM (1990) Acute flow-through toxicity of ^{14}C-cyfluthrin to *Daphnia magna*. Study number 100321. Analytical Bio-Chemistry Laboratories Inc. Columbia, MS. Submitted to U.S. Environmental Protection Agency. CDPR ID: 50317-135.

Carlisle JC (1984) Effects of cyfluthrin (technical Baythroid) on mallard duck reproduction study number 83-675-06. Mobay Chemical Corp report # 86690. Mobay Environmental Health Research Corporate Toxicology Dept., Stilwell, KS. Submitted to U.S. Environmental Protection Agency. CDPR ID: 50317-027.

Carlisle JC (1985) Toxicity of cyfluthrin (Baythroid) technical to early life stages of rainbow trout. Mobay Chemical Co. Study No. 85-666-01. Study number 90801. Mobay Chemical Corporation, Corporate Toxicology Dept. Environmental Health Research, Stilwell, KS. Submitted to U.S. Environmental Protection Agency. CDPR ID: 50317-090 and 50317-043.

Carlisle JC, Roney DJ (1984) Bioconcentration of cyfluthrin (Baythroid) by bluegill sunfish study number 83-766-01. Mobay Environmental Health Research Corporate Toxicology Dept. Stilwell, KS. Submitted to U.S. Environmental Protection Agency. Study number 86215. CDPR ID: 50317-006 and 50317-027.

CCME (2006) Canadian water quality guidelines: Permethrin. Scientific supporting document. Canadian Council of Ministers of the Environment, Winnipeg, Canada.

CDFG (2010a) State and federally listed endangered and threatened animals of California. California Natural Diversity Database. California Department of Fish and Game, Sacramento, CA. Available from: http://www.dfg.ca.gov/biogeodata/cnddb/pdfs/TEAnimals.pdf.

CDFG (2010b) State and federally listed endangered, threatened, and rare plants of California. California Natural Diversity Database. California Department of Fish and Game, Sacramento, CA. Available from: http://www.dfg.ca.gov/biogeodata/cnddb/pdfs/TEPlants.pdf.

Conrad AU, Fleming RJ, Crane M (1999) Laboratory and field response of *Chironomus riparius* to a pyrethroid insecticide. Water Res 33:1603–1610.

Corbel V, Chandre F, Darriet F, Lardeux F, Hougard J-M (2003) Synergism between permethrin and propoxur against *Culex quinquefasciatus* mosquito larvae. Medic Veterin Entomol 17:158–164.

Coulon (1982) Toxicity of Ambush® and Pydrin® to red crawfish, *Procambarus clarkii* (Girard) and channel catfish, *Ictalurus punctatus* Rafinesque in laboratory and field studies and the accumulation and dissipation of associated residues. Ph.D. Thesis Louisiana State University, Baton Rouge, LA.

Crommentuijn T, Sijm D, de Bruijn J, van Leeuwen K, van de Plassche E (2000) maximum permissible and negligible concentrations for some organic substances and pesticides. J Environ Manage 58:297–312.

Crossland NO (1982) Aquatic toxicology of cypermethrin. II. Fate and biological effects in pond experiments. Aquatic Toxicol 2:205–222.

CSIRO (2001) BurrliOZ, version 1.0.13. [cited 28 September 2010]. Commonwealth Scientific and Industrial Research Organization, Australia. Available from: http://www.cmis.csiro.au/Envir/burrlioz/.

Cutkomp LK, Subramanyam B (1986) Toxicity of pyrethroids to *Aedes aegypti* larvae in relation to temperature. J Am Mosquito Contr 2:347–349.

Day KE (1991) Effects of dissolved organic carbon on accumulation and acute toxicity of fenvalerate, deltamethrin and cyhalothrin to *Daphnia magna* (Straus). Environ Toxicol Chem 10:91–101.

DeLorenzo ME, Serrano L, Chung KW, Hoguet J, Key PB (2006) Effects of the insecticide permethrin on three life stages of the grass shrimp, *Palaemonetes pugio*. Ecotoxicol Environ Saf 64:122–127.

Drenner RW, Hoagland KD, Smith JD, Barcellona WJ, Johnson PC, Palmieri MA, Hobson JF (1993) Effects of sediment-bound bifenthrin on gizzard shad and plankton in experimental tank mesocosms. Environ Toxicol Chem 12:1297–1306.

Dwyer FJ, Hardesty DK, Henke CE, Ingersoll CG, Whites DW, Mount DR, Bridges CM (1999) Assessing contaminant sensitivity of endangered and threatened species: toxicant classes. EPA/600/R-99/098.

Dwyer FJ, Mayer FL, Sappington LC, Buckler DR, Bridges CM, Greer IE, Hardesty DK, Henke CE, Ingersoll CG, Kunz JL, Whites DW, Augspurger T, Mount DR, Hattala K, Neuderfer GN (2005) Assessing contaminant sensitivity of endangered and threatened aquatic species: Part I. Acute toxicity of five chemicals. Arch Environ Contam Toxicol 48:143–154.

Dwyer FJ, Sappington LC, Buckler DR, Jones SB (1995) Use of a surrogate species in assessing contaminant risk to endangered and threatened fishes. Final report – September, 1995. EPA/600/R-96/029.

Farmer D, Hill IR, Maund SJ (1995) A comparison of the fate and effects of two pyrethroid insecticides (lambda-cyhalothrin and cypermethrin) in pond mesocosms. Ecotoxicol 4:219–244.

Farrelly E, Hamer MJ (1989) PP321: *Daphnia magna* life-cycle study using a flow-through system. ICI Agrochemicals, Jealotts Hill Research Station. Bracknell, Berkshire, UK. Submitted to U.S. Environmental Protection Agency. USEPA MRID: 41217501, CDPR ID: 50907-089.

Forbis AD, Burgess D, Franklin L, Galbraith A (1984) Chronic toxicity of ^{14}C-cyfluthrin to *Daphnia magna* under flow-through conditions. Mobay Chemical Company. Analytical Bio-Chemistry Laboratories, Inc. Columbia, MO. Submitted to U.S. Environmental Protection Agency. Study number 88690. CDPR ID: 50317-090.

Friberg-Jensen U, Wendt-Rasch L, Woin P, Christoffersen K (2003) Effects of the pyrethroid insecticide, cypermethrin, on a freshwater community studied under field conditions. I. Direct and indirect effects on abundance measures of organisms at different trophic levels. Aquat Toxicol 63: 357–371.

Gagliano GG (1994) Acute toxicity of ^{14}C-cyfluthirn to the bluegill (*Lepomis macrochirus*) under flow-through conditions. Miles Incorporated Agriculture Division, Research and Development

Dept. Environmental Research Section, Stilwell, KS. Submitted to U.S. Environmental Protection Agency. USEPA MRID: 454267-07.

Gagliano GG, Bowers LM (1994) Acute toxicity of ^{14}C-cyfluthrin to the Rainbow trout (*Oncorhynchus mykiss*) under flow-through conditions. Miles Incorporated Agriculture Division, Research and Development Dept. Environmental Research Section, Stilwell, KS. Submitted to U.S. Environmental Protection Agency. USEPA MRID: 45426708.

Gan J, Lee SJ, Liu WP, Haver DL, Kabashima JN (2005) Distribution and persistence of pyrethroids in runoff sediments. J Environ Qual 34:836–841.

Gartenstein S, Quinnell RG, Larkum AWD (2006) Toxicity effects of diflubenzuron, cypermethrin and diazinon on the development of *Artemia salina* and *Heliocidaris tuberculata*. Australasian J Ecotoxicol 12:83–90.

Gomez-Gutierrez A, Jover E, Bayona JM, Albaiges J (2007) Influence of water filtration on the determination of a wide range of dissolved contaminants at parts-per-trillion levels. Anal Chim Acta 583:202–209.

Guay I, Roy MA, Samson R (2000) Recommandations de critères de qualité de l'eau pour la perméthrine pour la protection de la vie aquatique. Direction du suivi de l'état de l'environnement, Service des avis et des expertises, Ministère de l'Environnement du Québec, Québec.

Gunther U, Herrmann RA (1986) Baythroid pond study. Study number 91233. Mobay Corporation. OKOLIMNA Gesellschaft fur Okologie und Gewasserkunde mbH, Burgwedel, Germany. Submitted to California Department of Pesticide Regulation. CDPR ID: 50317-058.

Guy D (2000a) Aquatic Toxicology laboratory Report P-2161-2. Bifenthrin with cladoceran *Ceriodaphnia dubia* in an acute definitive test. California Department of Fish and Game, Aquatic Toxicology Lab, Elk Grove, CA.

Guy D (2000b) Aquatic Toxicology laboratory Report P-2161-2. Bifenthrin with *Pimephales promelas* in an acute definitive test. California Department of Fish and Game, Aquatic Toxicology Lab, Elk Grove, CA.

Hamer MJ (1997) Cypermethrin: Acute toxicity of short-term exposures to *Hyalella azteca*. Laboratory project ID: TMJ3904B. Zeneca Agrochemicals, Jealott's Hill Research Station, Bracknell, Berkshire, UK. EPA MRID: 44423501.

Hamer MJ, Ashwell JA, Gentle WE (1998) Lambda-cyhalothrin acute toxicity to aquatic arthropods. ZENECA Agrochemicals, Jealott's Hill Research Station, Bracknell, Berkshire, UK. Submitted to California Department of Pesticide Regulation. CDPR ID: 509-7-093.

Hamer MJ, Farrelly E, Hill IR (1985a) PP321: Toxicity to *Gammarus pulex*. ICI Plant Protection Division, Bracknell, Berkshire, UK. Submitted to California Department of Pesticide Regulation. CDPR ID: 509007-086.

Hamer MJ, Farrelly E, Hill IR (1985b) PP321: 21 day *Daphnia magna* life-cycle study. ICI Plant Protection Division, Bracknell, Berkshire, UK. Submitted to California Department of Pesticide Regulation. CDPR ID: 50907-089.

Hardstone MC, Leichter C, Harrington LC, Kasai S, Tomita T, Scott JG (2007) Cytochrome P450 monooxygenase-mediated permethrin resistance confers limited and larval specific cross-resistance in the southern house mosquito, *Culex pipiens quinquefasciatus*. Pestic Biochem Physiol 89:175–184.

Hardstone MC, Leichter C, Harrington LC, Kasai S, Tomita T, Scott JG (2008) Corrigendum to "Cytochrome P450 monooxygenase-mediated permethrin resistance confers limited and larval specific cross-resistance in the southern house mosquito, *Culex pipiens quinquefasciatus*." Pestic Biochem Physiol 91:191.

Harwood AD, You J, Lydy MJ (2009) Temperature as a toxicity identification evaluation tool for pyrethroid insecticides: Toxicokinetic confirmation. Environ Toxicol Contam 28:1051–1058.

Heath S, Bennett WA, Kennedy J, Beitinger TL (1994) Heat and cold tolerance of the fathead minnow, *Pimephales promelas*, exposed to the synthetic pyrethroid cyfluthrin. Can J Fish Aquat Sci 51:437–440.

Hill IR, Runnalls JK, Kennedy JH, Ekoniak P (1994) Lambda-cyhalothrin: A mesocosm study of its effects on aquatic organisms. In: Graney RL, Kennedy JH, Rodgers JH (eds) Aquatic Mesocosm Studies in Ecological Risk Assessment. Lewis Publishers, CRC Press, Boca Raton, FL.

Hill RW (1984a) PP321: Determination of acute toxicity to rainbow trout (*Salmo gairdneri*). ICI, Brixham laboratory, Brixham, Devon, UK. Submitted to California Department of Pesticide Regulation. CDPR ID: 50907-008.

Hill RW (1984b) PP321: Determination of acute toxicity to bluegill sunfish (*Lepomis macrochirus*). ICI, Brixham laboratory, Brixham, Devon, UK. Submitted to California Department of Pesticide Regulation. CDPR ID: 50907-085.

Hill RW, Caunter JE, Cumming RI (1985) PP321: Determination of the chronic toxicity to sheepshead minnow (*Cyprinodon variegatus*) embryos and larvae. ICI, Brixham laboratory, Brixham, Devon, UK. Submitted to California Department of Pesticide Regulation. CDPR ID: 50907-088.

Hladik ML, Kuivila KM (2009) Assessing the occurrence and distribution of pyrethroids in water and suspended sediments. J Agric Food Chem 57:9079–9085.

Hoagland KD, Drenner RW, Smith JD, Cross DR (1993) Freshwater community responses to mixtures of agricultural pesticides: effects of atrazine and bifenthrin. Environ Toxicol Chem 12:627–637.

Hoberg JR (1983a) Acute toxicity of FMC 54800 technical to bluegill (*Lepomis macrochirus*). FMC Study No: A83-987. EG &G, Bionomics study, Wareham, MA. Submitted to U.S. Environmental Protection Agency. EPA MRID: 00132536.

Hoberg JR (1983b) Acute toxicity of FMC 54800 technical to rainbow trout (*Salmo gairdneri*). FMC Study No: A83/967, Wareham, MA. Submitted to U.S. Environmental Protection Agency. EPA MRID: 00132539.

Holcombe GW, Phipps GL, Tanner DK (1982) The acute toxicity of Kelthane, Dursban, disulfoton, Pydrin, and permethrin to fathead minnows *Pimephales promelas* and rainbow trout *Salmo gairdneri*. Environ Pollut A 29:167–178.

Holmes RW, Anderson BS, Phillips BM, Hunt JW, Crane DB, Mekebri A, Connor V (2008) Statewide investigation of the role of pyrethroid pesticides in sediment toxicity in California's urban waterways. Environ Sci Technol 42:7003–7009.

Hunter W, Xu YP, Spurlock F, Gan J (2008) Using disposable polydimethylsiloxane fibers to assess the bioavailability of permethrin in sediment. Environ Toxicol Chem 27:568–575.

Johnson PC (1992) Impacts of the pyrethroid insecticide cyfluthrin on aquatic invertebrate populations in outdoor experimental tanks [dissertation]. University of North Texas, Denton, TX. Submitted to California Department of Pesticide Regulation. Study number 105036. CDPR ID: 50317-090.

Johnson PC, Kennedy JH, Morris RG, Hambleton FE (1994) Fate and effects of cyfluthrin (pyrethroid insecticide) in pond mesocosms and concrete microcosms. In: Graney RL, Kennedy JH, Rogers JH (eds) Aquatic Mesocosm Studies in Ecological Risk Assessment. CRC Press, Inc., Boca Raton, FL. p. 337–369.

Kasai S, Weerashinghe IS, Shono T (1998) P450 monoosygenases are an important mechanism of permethrin resistance in *Culex quinquefasciatus* Say larvae. Arch Insect Biochem Physiol 37:47–56.

Kennedy J, Johnson P, Montandon R (1990) Assessment of the potential ecological/biological effects of Baythroid® (cyfluthrin) utilizing artificial pond systems. Mobay Corporation Agricultural Chemical Division, University of North Texas, Kansas City, MO. Submitted to California Department of Pesticide Regulation. Study number 100147. CDPR ID: 50317-114.

Kent SJ, Shillabeer N (1997a) Lambda-cyhalothrin: Acute toxicity to golden orfe (*Leuciscus idus*). ZENECA Agrochemicals, Brixham Environmental Laboratory, Brixham, Devon, UK. Submitted to California Department of Pesticide Regulation. CDPR ID: 50907-085.

Kent SJ, Shillabeer N (1997b) Lambda-cyhalothrin: Acute toxicity to the guppy (*Poecilia reticulata*). ZENECA Agrochemicals, Brixham Environmental Laboratory, Brixham, Devon, UK. Submitted to California Department of Pesticide Regulation. CDPR ID: 50907-085.

Kent SJ, Shillabeer N (1997c) Lambda-cyhalothrin: Acute toxicity to zebra danio (*Brachydanio rerio*). ZENECA Agrochemicals, Brixham Environmental Laboratory, Brixham, Devon, UK. Submitted to California Department of Pesticide Regulation. CDPR ID: 50907-085.

Kent SJ, Shillabeer N (1997d) Lambda-cyhalothrin: Acute toxicity to fathead minnow (*Pimephales promelas*). ZENECA Agrochemicals, Brixham Environmental Laboratory, Brixham, Devon, UK. Submitted to California Department of Pesticide Regulation. CDPR ID: 50907-085.

Kent SJ, Williams NJ, Gillings E, Morris DS (1995a) Permethrin: chronic toxicity to *Daphnia magna*. Zeneca Brixham Environmental Laboratory, Brixham, UK. Laboratory project ID BL5443/B. EPA MRID 43745701.

Kim Y, Jung J, Oh S, Choi K (2008) Aquatic toxicity of cartap and cypermethrin to different life stages of *Daphnia magna* and *Oryzias latipes*. J Environ Sci Health B 43:56–64.

Kumaraguru AK, Beamish FWH (1981) Lethal toxicity of permethrin (NRDC-143) to rainbow trout, in relation to body-weight and water temperature. Water Res 15:503–505.

Lajmanovich R, Lorenzatti E, de la Sierra P, Marino F, Stringhini G, Peltzer P (2003) Reduction in the mortality of tadpoles (*Physalaemus biligonigerus*; Amphibia: Leptodactylidae) exposed to cypermethrin in presence of aquatic ferns. Fresenius Environ Bull 12:1558–1561.

Laskowski DA (2002) Physical and chemical properties of pyrethroids. Rev Environ Contam Toxicol 174:49–170.

Lauridsen RB, Friberg N (2005) Stream macroinvertebrate drift response to pulsed exposure of the synthetic pyrethroid lambda-cyhalothrin. Environ Toxicol 20:513–521.

LeBlanc GA (1976) Acute toxicity of FMC-33297 technical to *Daphnia magna*. EG&G, Bionomics, Wareham, MA. CDPR ID: study number 15100.

Lepper P, Sorokin N, Atkinson C, Rule K, Hope S-J, Comber S (2007) Proposed EQS for Water Framework Directive Annex VIII substances: permethrin. Science project number: SC040038. Product code: SCHO0407BLWF-E-E. Environment Agency, Bristol, UK. Available at: http://www.wfduk.org/LibraryPublicDocs/sr12007-permethrin.

Li H-P, Lin C-H, Jen J-F (2009) Analysis of aqueous pyrethroid residuals by one-step microwave-assisted headspace solid-phase microextraction and gas chromatography with electron capture detection. Talanta 79:466–471.

Long KWJ, Shillabeer N (1997a) Lambda-cyhalothrin: Acute toxicity to the three-spined stickleback (*Gasterosteus aculeatus*). ZENECA Agrochemicals, Brixham Environmental Laboratory, Brixham, Devon, UK. Submitted to California Department of Pesticide Regulation. CDPR ID: 50907-085.

Long KWJ, Shillabeer N (1997b) Lambda-cyhalothrin: Acute toxicity to channel catfish (*Ictalurus punctatus*). ZENECA Agrochemicals, Brixham Environmental Laboratory, Brixham, Devon, UK. Submitted to California Department of Pesticide Regulation. CDPR ID: 50907-085.

Lutnicka H, Bogacka T, Wolska L (1999) Degradation of pyrethroids in an aquatic ecosystem model. Water Res 33:3441–3446.

Machado MW (2001) XDE-225 and Lambda-cyhalothrin: comparative toxicity to Rainbow Trout (*Oncorhynchus mykiss*) under flow-through conditions. Springborn Laboratories, Inc. Wareham, MA. Submitted to U.S. Environmental Protection Agency. USEPA MRID: 45447217.

Mackay D, Shiu WY, Ma KC, Lee SC (2006) Handbook of Physical-Chemical Properties and Environmental Fate for Organic Chemicals, 2nd edn. CRC Press, Boca Raton, FL.

Marino TA, Rick DL (2001) XR-225 and Lambda-cyhalothrin: An acute toxicity comparison study with the Bluegill Sunfish, *Lepomis macrochirus* RAFINESQUE. Toxicology & Environmental Research and Consulting, The Dow Chemical Company, Midland, MI. Submitted to U.S. Environmental Protection Agency. USEPA MRID: 45447216.

Maund S, Biggs J, Williams P, Whitfield M, Sherratt T, Powley W, Heneghan P, Jepson P, Shillabeer N (2009) The influence of simulated immigration and chemical persistence on recovery of macroinvertebrates from cypermethrin and 3,4-dichloroaniline exposure in aquatic microcosms. Pest Manag Sci 65:678–687.

Mayer LM, Weston DP, Bock MJ (2001) Benzo[a]pyrene and zinc solubilization by digestive fluids of benthic invertebrates – A cross-phyletic study. Environ Toxicol Chem 20:1890–1900.

McAllister WA (1988) Full life cycle toxicity of ^{14}C-FMC 54800 to the fathead minnow (*Pimephales promelas*) in a flow-through system. FMC Study No: A86-2100. Analytical

Bio-Chemistry Laboratories, Inc. Columbia, MO. Submitted to U.S. Environmental Protection Agency. EPA MRID: 40791301.

Medina M, Barata C, Telfer T, Baird DJ (2004) Effects of cypermethrin on marine plankton communities: a simulated field study using mesocosms. Ecotoxicol Environ Saf 58:236–245.

Miller TA, Salgado VL (1985) The mode of action of pyrethroids on insects. In: Leahey JP (ed) The Pyrethroid Insecticides. Taylor & Francis, Philadelphia, PA.

Morris RG (1991) Pyrethroid insecticide effects on bluegill sunfish (*Lepomis macrochirus*) and the impacts of bluegill predation on invertebrates in microcosms. Mobay Corporation, University of North Texas, Denton, TX. Submitted to California Department of Pesticide Regulation. Study number 101953. CDPR ID: 50317-129.

Morris RG, Kennedy JH, Johnson PC, Hambleton FE (1994) Pyrethroid insecticide effects on bluegill sunfish in microcosm and mesocosm and bluegill impact on microcosm fauna. In: Graney RL, Kennedy JH, Rogers JH (eds) Aquatic Mesocosm Studies in Ecological Risk Assessment. CRC Press, Inc., Boca Raton, FL. pp 373–395.

Muir DCG, Hobden BR, Servos MR (1994) Bioconcentration of pyrethroid insecticides and DDT by rainbow trout: Uptake, depuration, and effect of dissolved organic carbon. Aquat Toxicol 29:223–240.

Muir DCG, Rawn GP, Townsend BE, Lockhart WL, Greenhalgh R (1985) Bioconcentration of cypermethrin, deltamethrin, fenvalerate and permethrin by *Chironomus tentans* larvae in sediment and water. Environ Toxicol Chem 4:51–61.

Narahashi T, Ginsburg KS, Nagata K, Song JH, Tatebayashi H (1998) Ion channels as targets for insecticides. Neurotoxicol 19:581–590.

Norgaard KB, Cedergreen N (2010) Pesticide cocktails can interact synergistically on aquatic crustaceans. Environ Sci Pollut Res 17:957–967.

Palmquist KR, Jenkins JJ, Jepson PC (2008) Effects of dietary esfenvalerate exposures on three aquatic insect species representing different functional feeding groups. Environ Toxicol Chem 27:1721–1727.

Palumbo AJ, TenBrook PL, Fojut TL, Faria IR, Tjeerdema RS (2012) Aquatic life water quality criteria derived via the UC Davis method: I. Organophosphate insecticides. Rev Environ Contam Toxicol 216:1–49.

Paul A, Harrington LC, Scott JG (2006) Evaluation of novel insecticides for control of Dengue vector *Aedes aegypti* (Diptera: Culicidae). J Med Entomol 43:55–60.

Paul EA, Simonin HA (2006) Toxicity of three mosquito insecticides to crayfish. Bull Environ Contam Toxicol 76:614–621.

Paul EA, Simonin HA, Tomajer TM (2005) A comparison of the toxicity of synergized and technical formulation of permethrin, sumithrin, and resmethrin to trout. Arch Environ Contam Toxicol 48:251–259.

Phillips BM, Anderson BS, Hunt JW, Nicely PA, Kosaka RA, Tjeerdema RS, de Vlaming V, Richard N (2004) In situ water and sediment toxicity in an agricultural watershed. Environ Toxicol Chem 23:435–442.

Phillips BM, Anderson BS, Hunt JW, Tjeerdema RS, Carpio-Obeso M, Connor V (2007) Causes of water toxicity to *Hyalella azteca* in the New River, California, USA. Environ Toxicol Chem 26:1074–1079.

Phillips BM, Anderson BS, Voorhees JP, Hunt JW, Holmes RW, Mekebri A, Connor V, Tjeerdema RS (2010) The contribution of pyrethroid pesticides to sediment toxicity in four urban creeks in California, USA. J Pestic Sci 35:302–309.

Poirier DG, Surgeoner GA (1988) Evaluation of a field bioassay technique to predict the impact of aerial applications of forestry insecticides on stream invertebrates. Can Ent 120:627–637.

Raimondo S, Vivian DN, Barron MG (2010) Web-based Interspecies Correlation Estimation (Web-ICE) for Acute Toxicity: User Manual. Version 3.1. Office of Research and Development, U.S. Environmental Protection Agency, Gulf Breeze, FL. EPA/600/R-10/004.

Rasmussen JJ, Friberg N, Larsen SE (2008) Impact of lambda-cyhalothrin on a macroinvertebrate assemblage in outdoor experimental channels: Implications for ecosystem functioning. Aquat Toxicol 90:228–234.

Rhodes JE, McAllister WA, Leak T, Stuerman L (1990) Full life-cycle toxicity of ^{14}C-cyfluthrin (Baythroid) to the fathead minnow (*Pimephales promelas*) under flow-through conditions. Mobay Corporation, Analytical Bio-Chemistry Laboratories, Inc. Aquatic Toxicology Division, Columbia, MO. Submitted to California Department of Pesticide Regulation. Study number 100097. CDPR ID: 50317-110.

Roberts NL, Phillips C, Anderson A, MacDonald I, Dawe IS, Chanter DO (1986) The effect of dietary inclusion of FMC 54800 on reproduction in the mallard duck. FMC Study No: A84/1260. FMC Corporation, Princeton, NJ. Submitted to U.S. Environmental Protection Agency. EPA MRID: 00163099.

Rodriguez MM, Bisset JA, de Armas Y, Ramos F (2005) Pyrethroid insecticide-resistant strain of *Aedes aegypti* from Cuba induced by deltamethrin selection. J Am Mosquito Cont Assoc 21:437–445.

Rodriguez MM, Bisset JA, Fernandez D (2007) Levels of insecticide resistance and resistance mechanisms in *Aedes aegypti* from some Latin American countries. J Am Mosquito Cont Assoc 23:420–429.

Roessink I, Arts GHP, Belgers JDM, Bransen F, Maund SJ, Brock TCM (2005) Effects of lambda-cyhalothrin in two ditch microcosm systems of different trophic status. Environ Toxicol Chem 24:1684–1696.

Sangster Research Laboratories (2010) LOGKOW. A databank of evaluated octanol-water partition coefficients (Log P). Canadian National Committee for CODATA. Available at: http://logkow.cisti.nrc.ca/logkow/index.jsp.

Sappington LC, Mayer FL, Dwyer FJ, Buckler DR, Jones JR, Ellersieck MR (2001) Contaminant sensitivity of threatened and endangered fishes compared to standard surrogate species. Environ Toxicol Chem 20:2869–2876.

Schroer AFW, Belgers JDM, Brock TCM, Matser AM, Maund SJ, Vann den Brink PJ (2004) Comparison of laboratory single species and field population-level effects of the pyrethroid insecticide λ-cyhalothrin on freshwater invertebrates. Arch Environ Contam Toxicol 46:324–335.

Semple KT, Doick KJ, Jones KC, Burauel P, Craven A, Harms H (2004) Defining bioavailability and bioaccessibility of contaminated soil and sediment is complicated. Environ Sci Technol 38:228A-231A.

Sherman JW (1989) Bifenthrin pond study: Ecological effects during treatment and post treatment follow-up studies of Hagan's pond, Orrville, Alabama. FMC report no. A84-1285-02 (January 25, 1989). Unpublished report prepared by the Academy of Natural Sciences of Philadelphia for FMC Corporation, Philadelphia, PA. Submitted to U.S. Environmental Protection Agency. EPA MRID: 40981822.

Siepmann S, Holm S (2000) Hazard assessment of the synthetic pyrethroid insecticides bifenthrin, cypermethrin, esfenvalerate, and permethrin to aquatic organisms in the Sacramento-San Joaquin River system. Administrative Report. California Department of Fish and Game, Rancho Cordova, CA.

Singh DK, Agarwal RA (1986) Piperonyl butoxide synergism with two synthetic pyrethroids against *Lymnaea acuminata*. Chemosphere 15:493–498.

Smith S, Lizotte RE (2007) Influence of selected water quality characteristics on the toxicity of λ-cyhalothrin and γ-cyhalothrin to *Hyalella azteca*. Bull Environ Contam Toxicol 79:548–551.

Spehar RL, Tanner DK, Nordling BR (1983) Toxicity of the Synthetic pyrethroids, permethrin and AC 222, 705 and their accumulation in early life stages of fathead minnows and snails. Aquat Toxicol 3:171–182

Stephenson RR (1982) Aquatic toxicology of cypermethrin I. Acute toxicity to some freshwater fish and invertebrates in laboratory tests. Aquatic Toxicol 2:175–185.

Stephenson RR, Choi SY, Olmos-Jerez A (1984) Determining the toxicity and hazard to fish of a rice insecticide. Crop Protection 3:151–165.

Stratton GW, Corke CT (1981) Interaction of permethrin with *Daphnia magna* in the presence and absence of particulate material. Environ Pollut A 24:135–144.

Sullivan K, Martin DJ, Cardwell RD, Toll JE, Duke S (2000) An analysis of the effects of temperature on salmonids of the Pacific Northwest with implications for selecting temperature criteria [internet]. [cited 2007 June 17]. Sustainable Ecosystems Institute, Portland, OR. Available from: http://www.sei.org.

Surprenant DC (1983) Acute toxicity of FMC 54800 technical to *Daphnia magna*. Bionomics Study. FMC Study No: A83-986. EG & G, Bionomics, Wareham, MA. Submitted to U.S. Environmental Protection Agency. EPA MRID: 00132537.

Surprenant DC (1986) Accumulation and elimination of ^{14}C-residues by bluegill (*Lepomis macrochirus*) exposed to ^{14}C-FMC 54800. FMC Study No: 182E54E01/85-4-176. Springborn Bionomics Inc. Wareham, MA. Submitted to U.S. Environmental Protection Agency. EPA MRID: 00163094; 470271-031.

Surprenant DC (1988) Bioavailability, accumulation and aquatic toxicity of ^{14}C-FMC 54800 residues incorporated into soil. FMC Study No: A85-1576. Springborn Bionomics Study No: 282-0185-6109-000, Wareham, MA. Submitted to U.S. Environmental Protection Agency. EPA MRID: 42529902.

Surprenant DC (1990) Acute toxicity of ^{14}C-®Baythroid to crayfish (*Procambarus clarkii*) under flow-through conditions. Mobay Corporation, Springborn Laboratories Inc. Wareham, MA. Submitted to California Department of Pesticide Regulation. Study number 100108. CDPR ID: 50317-112.

Tapp JF, Hill RW, Maddock BG, Harland BJ, Stembridge HM, Bolygo E (1988) Cypermethrin: Determination of chronic toxicity to fathead minnow (*Pimephales promelas*) full lifecycle. Laboratory project ID: BL/B/3173. ICI PLC, Brixham Laboratory, Brixham, Devon, UK. EPA MRID 40641701.

Tapp JF, Maddock BG, Harland BJ, Stembridge HM, Gillings E (1990) Lambda-cyhalothrin (Karate PP321): Determination of chronic toxicity to fathead minnow (*Pimephales promelas*) full lifecycle. Imperial Chemical Industries PLC, Brixham, Devon, UK. Submitted to U.S. Environmental Protection Agency. USEPA MRID: 41519001.

Tapp JF, Sankey SA, Caunter JE, Harland BJ (1989) Lambda-cyhalothrin: Determination of acute toxicity to Rainbow trout (*Salmo gairdneri*). Imperial Chemical Industries PLC, Brixham, Devon, UK. Submitted to California Department of Pesticide Regulation. CDPR ID: 50907-085.

TenBrook PL, Palumbo AJ, Fojut TL, Hann P, Karkoski J, Tjeerdema RS (2010) The University of California-Davis Methodology for deriving aquatic life pesticide water quality criteria. Rev Environ Contam Toxicol 209:1–155.

TenBrook PL, Tjeerdema RS, Hann P, Karkoski J (2009) Methods for deriving pesticide aquatic life criteria. Rev Environ Contam Toxicol 199:19–109.

Thompson RS (1985) PP321: Determination of the acute toxicity to larvae of the Pacific oyster (*Crassostrea gigas*). Imperial Chemical Industries PLC, Brixham, Devon, UK. Submitted to California Department of Pesticide Regulation. CDPR study 50907-087.

Thompson RS (1986) Supplemental data in support of MRID 42584001. Permethrin: Determination of acute toxicity to mysid shrimps (*Mysidopsis bahia*). Laboratory project ID BL/B/2921. Brixham study no P131/B. Study performed by Brixham Environmental Laboratory, Devon, UK. EPA MRID 43492902.

Thompson RS, Williams TD, Tapp JF (1989) Permethrin: Determination of chronic toxicity to mysid shrimps (*Mysidopsis bahia*) (Run 2). Laboratory project ID: BL/B/3574. Study performed by Imperial Chemical Industries PLC Brixham Laboratory Freshwater Quarry: Brixham, Devon, UK. EPA MRID 41315701.

Tomlin CDS (ed) (2003) The Pesticide Manual, a World Compendium, 13th edition. British Crop Protection Council, Alton, Hampshire, UK.

Trimble AJ, Weston DP, Belden JB, Lydy MJ (2009) Identification and evaluation of pyrethroid insecticide mixtures in urban sediments. Environ Toxicol Chem 28:1687–1695.

UKTAG (2008) Proposals for Environmental Quality Standards for Annex VIII Substances. UK Technical Advisory Group on the Water Framework Directive. Available at: http://www.

wfduk.org/stakeholder_reviews/stakeholder_review_1-2007/LibraryPublicDocs/final_specific_pollutants.
USEPA (1985) Guidelines for deriving numerical national water quality criteria for the protection of aquatic organisms and their uses. U.S. Environmental Protection Agency, Springfield, VA. PB-85-227049.
USEPA (1996a) Ecological Effects Test Guidelines OPPTS 850.1010 Aquatic invertebrate acute toxicity test, freshwater daphnids. U.S. Environmental Protection Agency, Washington, DC. EPA 712–C–96–114.
USEPA (1996b) Ecological Effects Test Guidelines OPPTS 850.1045 Penaeid Acute Toxicity Test. U.S. Environmental Protection Agency, Washington, DC. EPA 712–C–96–137.
USEPA (2000) Methods for measuring the toxicity and bioaccumulation of sediment-associated contaminants with freshwater invertebrates. Second edition. U.S. Environmental Protection Agency, Washington, DC. EPA 600/R-99/064.
USEPA (2008) Reregistration eligibility decision for cypermethrin (revised 01/14/08). EPA OPP-2005-0293.
USFDA (2000) Industry activities staff booklet. U.S. Food and Drug Administration, Washington, DC. Available from: www.cfsan.fda.gov/~lrd/fdaact.html.
USFWS (2010) Species Reports. Endangered Species Program. U.S. Fish and Wildlife Service. Available from: http://www.fws.gov/endangered/; http://ecos.fws.gov/tess_public/pub/listedAnimals.jsp; http://ecos.fws.gov/tess_public/pub/listedPlants.jsp
Vaishnav DD, Yurk JJ (1990) Cypermethrin (FMC 45806): Acute toxicity to rainbow trout (*Oncorhynchus mykiss*) under flow-through test conditions. FMC Corporation study number A89-3109-01. Laboratory project ID: ESE No. 3903026-0750-3140. Environmental Science and Engineering, Inc. (ESE), Gainesville, FL. CDPR 118785.
Van Wijngaarden RPA, Brock TCM, Van Den Brink PJ, Gylstra R, Maund SJ (2006) Ecological effects of spring and late summer applications of lambda-cyhalothrin on freshwater microcosms. Arch Environ Contam Toxicol 50:220–239.
Ward GS (1986a) Acute toxicity of FMC 54800 technical on new shell growth of the eastern oyster (*Crassostrea virginica*). Environmental Science and Engineering, Inc. FMC Study No: A86-2083, Gainesville, FL. Submitted to U.S. Environmental Protection Agency. USEPA MRIDs 470271-040 and 00163103.
Ward GS (1986b) Acute toxicity of FMC 54800 technical on new shell growth of the eastern oyster (*Crassostrea virginica*). FMC Study No: A86-2203, Environmental Science and Engineering, Inc. Gainesville, FL. Submitted to U.S. Environmental Protection Agency. USEPA MRID 40266501.
Ward GS (1987) Acute toxicity of FMC 54800 technical to embryos and larvae of the eastern oyster (*Crassostrea virginica*). Environmental Science and Engineering, Inc. FMC Study No: A87-2264, Gainesville, FL. Submitted to U.S. Environmental Protection Agency. USEPA MRID 40383501.
Ward TJ, Boeri RL (1991) Acute toxicity of FMC 56701 technical and cypermethrin technical to daphnid, *Daphnia magna*. FMC Study: A90-3310. EnviroSystems Division, Hampton, NH. CDPR ID: 118786.
Wendt-Rasch L, Friberg-Jensen U, Woin P, Christoffersen K (2003) Effects of the pyrethroid insecticide cypermethrin on a freshwater community studied under field conditions. II. Direct and indirect effects on the species composition. Aquat Toxicol 63:373–389.
Wendt-Rasch L, Van den Brink PJ, Crum SJH, Woin P (2004) The effects of a pesticide mixture on aquatic ecosystems differing in trophic status: responses of the macrophyte *Myriophyllum spicatum* and the periphytic algal community. Ecotoxicol Environ Saf 57:383–398.
Werner RA, Hilgert JW (1992) Effects of permethrin on aquatic organisms in a freshwater stream in south-central Alaska. J Econ Entomol 85:860–864.
Werner I, Moran K (2008) Effects of pyrethroid insecticides on aquatic organisms. In Gan J, Spurlock F, Hendley P, Weston D (eds) Synthetic Pyrethroids: Occurrence and Behavior in Aquatic Environments. American Chemical Society, Washington, DC.

Weston DP, Amweg El, Mekebri A, Ogle RS, Lydy MJ (2006) Aquatic effects of aerial spraying for mosquito control over an urban area. Environ Sci Technol 40:5817–5822.

Weston DP, Holmes RW, Lydy MJ (2009a) Residential runoff as a source of pyrethroid pesticides to urban creeks. Environ Pollut 157:287–294.

Weston DP, Holmes RW, You J, Lydy MJ (2005) Aquatic toxicity due to residential use of pyrethroid insecticides. Environ Sci Technol 39:9778–9784.

Weston DP, Jackson CJ (2009) Use of engineered enzymes to identify organophosphate and pyrethroid-related toxicity in toxicity identification evaluations. Environ Sci Technol 43:5514–5520.

Weston DP, Lydy MJ (2010) Urban and agricultural sources of pyrethroid insecticides to the Sacramento-San Joaquin Delta of California. Environ Sci Technol 44:1833–1840.

Weston DP, You J, Harwood AD, Lydy MJ (2009b) Whole sediment toxicity identification evaluation tools for pyrethroid insecticides: III. Temperature manipulation. Environ Toxicol Chem 28:173–180.

Weston DP, You J, Lydy MJ (2004) Distribution and toxicity of sediment-associated pesticides in agriculture-dominated water bodies of California's Central Valley. Environ Sci Technol 38:2752–2759.

Weston DP, Zhang MH, Lydy MJ (2008) Identifying the cause and source of sediment toxicity in an agriculture-influenced creek. Environ Toxicol Chem 27:953–962.

Wheat J, Evans J (1994) Zetacypermethrin technical and cypermethrin technical: Comparative acute toxicity to the water flea, *Daphnia magna,* under flow-through conditions. FMC Study No. A92-3636. Laboratory project ID: J9210001b. Toxikon Environmental Sciences, Jupiter, FL. EPA MRID 432935-01.

Wheelock CE, Miller JL, Miller MJ, Gee SJ, Shan G, Hammock BD (2004) Development of toxicity identification evaluation procedures for pyrethroid detection using esterase activity. Environ Toxicol Chem 23:2699–2708.

Xu Q, Liu H, Zhang L, Liu N (2005) Resistance in the mosquito, *Culex quinquefasciatus*, and possible mechanisms for resistance. Pest Manag Sci 61:1096–1102.

Xu YP, Spurlock F, Wang ZJ, Gan J (2007) Comparison of five methods for measuring sediment toxicity of hydrophobic contaminants. Environ Sci Technol 41:8394–8399.

Yamauchi F, Shigeoka T, Yamagata T, Saito H, Suzuki Y (1984) PP-563 (Cyhalothrin): Accumulation in fish (Carp) in a flow-through water system. ICI Japan Limited, Mitsubishi-Kasei Institute of Toxicological and Environmental Sciences, Yokohama, Kanagawa, Japan. Submitted to California Department of Pesticide Regulation. CDPR ID: 50907-090.

Yang WC, Gan JY, Hunter W, Spurlock F (2006a) Effect of suspended solids on bioavailability of pyrethroid insecticides. Environ Toxicol Chem 25:1585–1591.

Yang WC, Hunter W, Spurlock F, Gan J (2007) Bioavailability of permethrin and cyfluthrin in surface waters with low levels of dissolved organic matter. J Environ Qual 36:1678–1685.

Yang WC, Spurlock F, Liu WP, Gan. JY (2006b) Inhibition of aquatic toxicity of pyrethroid insecticides by suspended sediment. Environ Toxicol Chem 25:1913–1919.

Yang WC, Spurlock F, Liu WP, Gan JY (2006c) Effects of dissolved organic matter on permethrin bioavailability to *Daphnia* species. J Agric Food Chem 54:3967–3972.

Yasuno M, Hanazato T, Iwakuma T, Takamura K, Ueno R, Takamura N (1988) Effects of permethrin on phytoplankton and zooplankton in an enclosure ecosystem in a pond. Hydrobiologia 159:247–258.

You J, Harwood AD, Li H, Lydy MJ (2011) Chemical techniques for assessing bioavailability of sediment-associated contaminants: SPME versus Tenax extraction. J Environ Monit 13:792–800.

Zhang Z-Y, Yu X-Y, Wang D-L, Yan H-J, Liu X-J (2010) Acute toxicity to zebrafish of two organophosphates and four pyrethroids and their binary mixtures. Pest Manag Sci 66:84–89.

Aquatic Life Water Quality Criteria Derived via the UC Davis Method: III. Diuron

Tessa L. Fojut, Amanda J. Palumbo, and Ronald S. Tjeerdema

1 Introduction

Diuron is a phenylurea herbicide that has been frequently detected in surface waters (the US Environmental Protection Agency, USEPA 2003), including periods when relatively low amounts were used, because it is moderately persistent in the water column (Ensminger et al. 2008). Diuron poses a risk to aquatic life because it, and other herbicides, can cause adverse effects on algae and vascular plants, which are the foundation of the aquatic food chain. Water quality standards are used to regulate pesticides in surface waters, and these standards are typically based on water quality criteria for the protection of aquatic life. When pesticide concentrations do not exceed water quality criteria, no adverse effects on aquatic life are expected. The derivation of acute and chronic water quality criteria for diuron using a new methodology developed by the University of California, Davis (TenBrook et al. 2010), is described in this chapter. The UC Davis methodology (UCDM) was designed to be more flexible than the USEPA method (1985) for deriving water quality criteria, although many aspects of the methods are similar.

2 Data Collection and Evaluation

Diuron (*N′*-(3,4-dichlorophenyl)-*N*, *N*-dimethylurea) is a phenylurea herbicide that is moderately soluble in water. Based on its physical–chemical properties, the herbicide is not likely to partition to sediments or to volatilize (Table 1), and it is considered to be moderately persistent because it is stable to hydrolysis (Table 2).

T.L. Fojut (✉) • A.J. Palumbo • R.S. Tjeerdema
Department of Environmental Toxicology, College of Agricultural and Environmental Sciences, University of California, Davis, CA 95616-8588, USA
e-mail: tlfojut@ucdavis.edu

R.S. Tjeerdema (ed.), *Aquatic Life Water Quality Criteria for Selected Pesticides*, Reviews of Environmental Contamination and Toxicology 216,
DOI 10.1007/978-1-4614-2260-0_3,

Table 1 Physical–chemical properties of diuron

Molecular weight	233.10
Density	1.4 g/mL (IUPAC 2008)
Water solubility	38 mg/L (geomean, $n = 2$; Tomlin 2003; IUPAC 2008)
Melting point	158°C (Lide 2003)
Vapor pressure	1.15×10^{-3} mPa (IUPAC 2008)
Henry's constant (K_H)	173,205 Pa m^3 mol^{-1} (geomean, $n = 2$; Mackay et al. 2006; IUPAC 2008)
Log K_{oc}[a]	2.61 (geomean, $n = 20$; Mackay et al. 2006)
Log K_{ow}[b]	2.78 (geomean, $n = 3$; Hansch et al. 1995; Sangster Research Laboratories 2008; IUPAC 2008)

[a] Log-normalized organic carbon–water partition coefficient
[b] Log-normalized octanol–water partition coefficient

Table 2 Environmental fate of diuron

	Half-life	Water	Temp (°C)	pH	Reference
Hydrolysis	>4 months	Phosphate buffer	20	5–9	Mackay et al. (2006)
	Stable	Sterile buffer	25	5, 7, 9	USEPA (2003)
Aqueous photolysis	2.25 h	Distilled	NR	NR	Mackay et al. (2006)
	43 days	NR	NR	NR	USEPA (2003)
Biodegradation (aerobic)	~20 days	Filtered sewage water	20	NR	Mackay et al. (2006)

NR not reported

Approximately 86 original studies on the effects of diuron on aquatic life were identified and reviewed. These studies are available in the open literature or may be requested from the USEPA or the California Department of Pesticide Regulation (CDPR). Studies that fell into three categories were evaluated according to the UCDM: (1) single-species effects, (2) ecosystem-level studies, and (3) terrestrial wildlife studies.

According to the UCDM scheme, single-species effect studies were rated for relevance and reliability, in a manner which was summarized by Palumbo et al. (2012). Studies that were rated as relevant (R) or less relevant (L) were also rated for reliability, whereas those that were rated as not relevant (N) were not further rated. There were three categories of study reliability: reliable (R), less reliable (L), or not reliable (N). The reliability ratings were determined by how many test parameters (e.g., nominal concentrations, source of dilution water, etc.) were reported, and if they were acceptable according to standard methods. Studies were then assigned a two-letter code in which their degree of relevance and reliability were rated. Studies that were rated not relevant (N) or not reliable (RN or LN) were not used for criteria derivation. All data rated as acceptable (RR) or supplemental (RL, LR, LL) for criteria derivation are summarized in Tables 3–7. Acceptable data rated as relevant and reliable (RR) were used for numeric criteria derivation. Supplemental data that were rated as less relevant and/or less reliable (RL, LR, or LL) for particularly sensitive, threatened, or endangered species were compared to the criteria to ensure protection of these species. Data summary records

Table 3 Final acute toxicity data set for diuron

Species	Test type	Meas/ Nom	Chemical grade (%)	Duration (h)	Temp (°C)	End point	Age/size	LC/EC_{50} (μg/L) (95% CI)	Reference
Daphnia magna	S	Nom	80.0	48	19.9	Mortality/ immobility	<24 h	12,000 (10,000–13,000)	Baer (1991)
Daphnia pulex	SR	Meas	99.8	96	22	Mortality	5 days	17,900 (14,200–22,600)	Nebeker and Schuytema (1998)
Hyalella azteca	SR	Meas	99.8	96	22	Mortality	<11 days	19,400 (17,700–21,300)	Nebeker and Schuytema (1998)

All studies were rated RR (data rated as acceptable)

S Static, *SR* static renewal, *FT* flow through

Table 4 Final chronic plant toxicity data set for diuron

Species	Test type	Meas/ Nom	Chemical grade (%)	Duration	Temp (°C)	End point	Age/size	NOEC[a] (µg/L)	LOEC[b] (µg/L)	MATC[c] (µg/L)	EC_{50} (µg/L) (95% CI)	Reference
Lemna gibba G3	S	Meas	99.1	7 days	24.7	Growth inhibition (biomass yield), relative growth rate (biomass)	Plant with 4 fronds	2.47	8.11	**4.48**	14.4 (9.26–19.6)[d]	Ferrell (2006)
Navicula pelliculosa	S	Nom	99.1	72 h	22–24	Growth inhibition (biomass)	Cells	11	33	19.1	22 (9–56)	Dengler (2006b)
N. pelliculosa	S	Nom	99.1	72 h	22–24	Growth inhibition (growth rate)	Cells	11	33	19.1	65 (33–160)	Dengler (2006b)
N. pelliculosa						Growth inhibition				**19.1**		Geomean
Pseudokirchneriella subcapitata (formerly *Selenastrum capricornutum* Printz)	S	Meas	96.8	120 h	24	Growth inhibition	2-day-old algal cells	1.3	2.5	**1.8**	2.9 (2.5–3.5)	Blasberg et al. (1991)
Scenedesmus obliquus	S	Nom	Technical	24 h	21	Growth inhibition	Algal cells	NR	NR	NR	10	Geoffroy et al. (2002)
Synechococcus leopoliensis	S	Nom	99.1	72 h	22–25	Growth inhibition (biomass)	Algal cells	3.7	11	**6.4**	26 (4–100)	Dengler (2006a)

All studies were rated RR

S Static, *SR* static renewal, *FT* flow through, *NR* not reported, *n/a* not applicable

Species mean chronic value is in bold

[a] No-observed effect concentration

[b] Lowest-observed effect concentration

[c] Maximum acceptable toxicant concentration

[d] EC_{50} based on biomass yield end point

Table 5 Final chronic animal toxicity data set for diuron

Species	Test type	Meas/ Nom	Chemical grade (%)	Duration (days)	Temp (°C)	End point	Age/size	NOEC (μg/L)	LOEC (μg/L)	MATC (μg/L)	Reference
Chironomus tentans	SR	Meas	99.8	10	24	Mortality	2 days, first instar	1,900	3,400	2,540	Nebeker and Schuytema (1998)
Daphnia pulex	S	Meas	99.8	7	NR	Reduced number of young/mortality	5 days	4,000.0	7,700	5,550	Nebeker and Schuytema (1998)
Hyalella azteca	SR	Meas	99.8	10	22	Mortality/reduced weight	<11 days	7,900	15,700	11,140	Nebeker and Schuytema (1998)
Lumbriculus variegatus	SR	Meas	99.8	10	23	Reduced weight	Small, short adults	1,800	3,500	2,510	Nebeker and Schuytema (1998)
Physa gyrina	SR	Meas	99.8	10	24	Reduced weight	2 days, first instar	13,400	22,800	17,480	Nebeker and Schuytema (1998)
Pimephales promelas	FT	Meas	98.6	64	25	Deformity, mortality	Eggs <24 h, hatched fry	33.4	78	51	Call et al. (1983, 1987)
Pseudacris regilla	SR	Meas	99.8	14	20		Tadpole	14,500	21,100	17,490	

(continued)

Table 5 (continued)

Species	Test type	Meas/Nom	Chemical grade (%)	Duration (days)	Temp (°C)	End point	Age/size	NOEC (μg/L)	LOEC (μg/L)	MATC (μg/L)	Reference
						Growth inhibition (length)					Schuytema and Nebeker (1998)
Rana aurora	SR	Meas	99.8	7	20	Growth inhibition (wet weight)	Tadpole	7,600	14,500	10,500	Schuytema and Nebeker (1998)
Rana catesbeiana	SR	Meas	99.8	21	24	Growth inhibition (dry weight)	Tadpole	11,690[a]	16,430[a]	12,450[a]	Schuytema and Nebeker (1998)
Xenopus laevis	SR	Meas	99.8	4 days	24	Growth inhibition (length)	Embryo	10,490[b]	20,540[b]	14,680[b]	Schuytema and Nebeker (1998)

All studies were rated RR

S static, *SR* static renewal, *FT* flow through, *NR* not reported

[a] SMCV calculated from three values

[b] SMCV calculated from two values

Table 6 Acceptable excluded data rated RR with given reason for exclusion

Species	Test type	Meas/ Nom	Chemical grade (%)	Duration	Temp (°C)	End point	Age/size	LC/EC$_{50}$ (μg/L) (95% CI)	MATC (μg/L)	Reference	Reason for exclusion
Chironomus tentans	SR	Meas	99.8	10 days	24	Reduced weight	2 days, first instar	–	4,910	Nebeker and Schuytema (1998)	A
Daphnia magna	S	Nom	80.0	24 h	19.9	Mortality/ immobility	<24 h	68,000 (55,000–86,000)	–	Baer (1991)	D
Lemna gibba G3	S	Meas	99.1	7 days	24.7	Growth inhibition (biomass)	Plant with 4 fronds	15.7 (10.06–20.8)	4.48	Ferrell (2006)	A
L. gibba G3	S	Meas	99.1	7 days	24.7	Growth inhibition (frond count)	Plant with 4 fronds	19.1 (13.4–24.8)	14.47	Ferrell (2006)	A
L. gibba G3	S	Meas	99.1	7 days	24.7	Growth inhibition (frond count yield)	Plant with 4 fronds	17.5 (11.8–23.2)	14.47	Ferrell (2006)	A
L. gibba G3	S	Meas	99.1	7 days	24.7	Relative growth rate (frond count)	Plant with 4 fronds	–	14.5	Ferrell (2006)	A
P. promelas	SR	Meas	99.8	7 days	25	Reduced weight	2.5 days embryo	–	5,900	Nebeker and Schuytema (1998)	C
P. promelas	SR	Meas	99.8	10 days	24	Mortality	1.5 months juvenile	–	23,280	Nebeker and Schuytema (1998)	B
Pseudacris regilla	SR	Meas	99.8	10 days	20	Increased deformity	Embryo	–	20,540	Schuytema and Nebeker (1998)	A

(continued)

Table 6 (continued)

Species	Test type	Meas/ Nom	Chemical grade (%)	Duration	Temp (°C)	End point	Age/size	LC/EC_{50} (μg/L) (95% CI)	MATC (μg/L)	Reference	Reason for exclusion
P. regilla	SR	Meas	99.8	14 days	20	Growth inhibition (wet weight)	Tadpole	–	24,720	Schuytema and Nebeker (1998)	A
P. regilla	SR	Meas	99.8	14 days	20	Growth inhibition (dry weight)	Tadpole	–	24,750[a]	Schuytema and Nebeker (1998)	A
Rana catesbeiana	SR	Meas	99.8	21 days	24	Growth inhibition (length)	Tadpole	–	18,950[a]	Schuytema and Nebeker (1998)	A
R. catesbeiana	SR	Meas	99.8	21 days	24	Growth inhibition (wet weight)	Tadpole	–	22,560[a]	Schuytema and Nebeker (1998)	A
Synechococcus leopoliensis	S	Nom	99.1	72 h	22–25	Growth inhibition (growth rate)	Algal cells	–	19.1	Dengler (2006a)	A
Xenopus laevis	SR	Meas	99.8	4 days	24	Deformity	Embryo	–	22,560	Schuytema and Nebeker (1998)	A

Reasons for exclusion
A. Less-sensitive end point
B. Less-sensitive life stage
C. Test type not preferred (static vs. flow through)
D. Not the most sensitive or appropriate duration
[a] SMCV calculated from two values

Table 7 Supplemental diuron data rated RL, LR, LL with given reason for rating and exclusion (listed at the end of table)

Species	Test type	Meas/ Nom	Chemical grade (%)	Duration	Temp (°C)	End point	Age/size	LC/EC_{50} (µg/L) (95% CI)	MATC (µg/L)	Reference	Rating/ reason
Achnanthes brevipes	S	Nom	Technical	3 days	20	Reduced oxygen evolution	Algal cells	24 (SE = 1.0)	–	Hollister and Walsh (1973)	LL/1, 2, 6
Americamysis bahia	FT	Meas	96.8	28 days	25.3	Number of young surviving	<24 h juvenile	–	1,400	Ward and Boeri (1992b)	RL/2
Amphora exigua	S	Nom	Technical	3 days	20	Reduced oxygen evolution	Algal cells	31 (SE = 4)	–	Hollister and Walsh (1973)	LL/1, 2, 6
Apium nodiflorum	S	Nom	>99	14 days	NR	Relative growth rate	Single stem node with leaf	2.808	NOEC = 0.05	Lambert et al. (2006)	LL/1, 5, 6
A. nodiflorum	S	Nom	>99	14 days	NR	Growth inhibition (roots)[b]	Single stem node with leaf	0.00026	NOEC < 0.0005	Lambert et al. (2006)	LL/1, 5, 6, 7
A. nodiflorum	S	Nom	>99	14 days	NR	Change in chlorophyll fluorescence ratio	Single stem node with leaf	>5.0	NOEC = 5	Lambert et al. (2006)	LL/1, 5, 6
Artemia salina	S	NR	NR	24 h	25	Mortality	Instar II–III larvae	12,010 (11,420–12,100)	–	Koutsaftis and Aoyama (2007)	LL/2, 5
Asellus brevicaudus	S	Nom	95.0	96 h	15	Mortality	Mature	15,500 (7,200–33,400)	–	Johnson and Finley (1980)	LL/5, 6
Chara vulgaris	S	Nom	>99	14 days	NR	Relative growth rate	Terminal lengths of shoots with 3 nodes	0.35	NOEC = 0.0005	Lambert et al. (2006)	LL/1, 5, 6

(continued)

Table 7 (continued)

Species	Test type	Meas/ Nom	Chemical grade (%)	Duration	Temp (°C)	End point	Age/size	LC/EC$_{50}$ (μg/L) (95% CI)	MATC (μg/L)	Reference	Rating/ reason
C. vulgaris	S	Nom	>99	14 days	NR	Change in chlorophyll fluorescence ratio	Terminal lengths of shoots with 3 nodes	4.033	NOEC = 0.5	Lambert et al. (2006)	LL/1, 5, 6
Chlamydomonas moewusii Gerloff	S	Nom	80.0	7 days	21	Growth inhibition	7-day-old algal cell stock	559.44	–	Cain and Cain (1983)	RL/1, 6
Chlamydomonas sp.	S	Nom	Technical	3 days	20	Reduced oxygen evolution	Algal cells	37 (SE = 3)	–	Hollister and Walsh (1973)	LL/1, 2, 6
Chlamydomonas sp.	S	Nom	99.8	20 min	21.5	Change in chlorophyll fluorescence ratio	2–4-week-old algal cells	10.8 (8.5–13.6)	0.22	Podola and Melkonian (2005)	RL/1, 5, 8
Chlorella pyrenoidosa	S	Nom	95.0	4 days	25	Growth inhibition	Algal cells	25	–	Maule and Wright (1984)	LR/1, 6
C. pyrenoidosa	S	Nom	50.0	96 h	25	Growth inhibition	Algal cells	1.3	–	Ma et al. (2001), Ma (2002)	LL/1, 3, 6
Chlorella sp.	S	Nom	Technical	10 days	20.5	Growth inhibition	Algal cells	EC_{66} = 4	–	Ukeles (1962)	LL/1, 2, 6
Chlorella sp.	S	Nom	Technical	3 days	20	Reduced oxygen evolution	Algal cells	19 (SE = 2)	–	Hollister and Walsh (1973)	LL/1, 2, 6

Chlorella vulgaris	S	Nom	50.0	96 h	25	Growth inhibition	Algal cells	4.3	–	Ma (2002)	LL/1, 3, 6
C. vulgaris SAG211-11b	S	Nom	99.8	20 min	21.5	Change in chlorophyll fluorescence ratio	2–4-week-old algal cells	27.4 (21.1–35.5)	0.22	Podola and Melkonian (2005)	RL/1, 8
Chlorococcum sp.	S	Nom	Technical	7 days	20	Growth inhibition	Algal cells	$EC_{62} = 10$	NOEC < 1.0	Walsh and Grow (1971)	RL/1, 2
Chlorococcum sp.	S	Nom	Technical	10 days	20	Growth inhibition	Algal cells	10	–	Walsh (1972)	RL/1, 2
Chlorococcum sp.	S	Nom	Technical	90 min	20	Reduced oxygen evolution	Algal cells	20	–	Walsh (1972)	RL/1, 2
Chlorococcum sp.	S	Nom	Technical	3 days	20	Reduced oxygen evolution	Algal cells	20 (SE = 4)	–	Hollister and Walsh (1973)	LL/1, 2, 6
Crassostrea virginica	FT	Meas	96.8	96 h	23	Shell deposition	Neonates, <24 h	4,800 (4,400–5,200)	NOEC = 2,400	Ward and Boeri (1991)	RL/2
Cryptomonas sp.	S	Nom	99.8	20 min	21.5	Change in chlorophyll fluorescence ratio	2–4-week-old algal cells	6.4 (5.3–7.8)	0.22	Podola and Melkonian (2005)	RL/1, 5, 8
Ctenopharyngodon idella	FT	NR	100.0	96 h	13	Mortality	1+ year, 15.8 g, 9.5 cm	31,000 (28,000–34,000)	–	Tooby et al. (1980)	LL/1, 5, 6
Cyclotella nana	S	Nom	Technical	3 days	20	Reduced oxygen evolution	Algal cells	39 (SE = 7)	–	Hollister and Walsh (1973)	LL/1, 2, 6
Cyprinodon variegates	FT	Meas	96.8	32 days	30	Mortality	<24 h	–	2,500	Ward and Boeri (1992a)	RL/2

(continued)

Table 7 (continued)

Species	Test type	Meas/ Nom	Chemical grade (%)	Duration	Temp (°C)	End point	Age/size	LC/EC$_{50}$ (μg/ L) (95% CI)	MATC (μg/L)	Reference	Rating/ reason
Daphnia magna	S	Nom	Technical	26 h	21.1	Mortality/ immobility	First instar	47,000 (41,600 –53,100)	–	Crosby and Tucker (1966)	LL/1, 5, 6
Daphnia pulex	S	Nom	95.0	48 h	15	Mortality/ immobility	First instar	1,400 (1,000–1,900)	–	Johnson and Finley (1980)	LL/5, 6
Dunaliella euchlora Lerche	S	Nom	Technical	10 days	20.5	Growth inhibition	Algal cells	$EC_{56} = 0.4$	–	Ukeles (1962)	LL/1, 2, 6
Dunaliella tertiolecta	S	Nom	99.0	96 h	20	Growth inhibition	Algal cells	5.9	–	Gatidou and Thomaidis (2007)	LL/2, 5
D. tertiolecta	S	Nom	Technical	3 days	20	Reduced oxygen evolution	Algal cells	10 (SE = 3)	–	Hollister and Walsh (1973)	LL/1, 2, 6
D. tertiolecta Butcher	S	Nom	Technical	10 days	20	Growth inhibition	Algal cells	20	–	Walsh (1972)	RL/1, 2
D. tertiolecta Butcher	S	Nom	Technical	90 min	20	Reduced oxygen evolution	Algal cells	10	–	Walsh (1972)	RL/2, 6, 8
Eudorina elegans	S	Nom	99.8	20 min	21.5	Change in chlorophyll fluorescence ratio	2–4-week-old algal cells	13.2 (10.4–16.9)	0.22	Podola and Melkonian (2005)	RL/1, 5, 8
Gammarus fasciatus	S	Nom	Technical	24 h	15.5	Mortality	Early instar	2,500 (1,000–5,500)	–	Sanders (1970)	LL/1, 5, 6
G. fasciatus	S	Nom	Technical	48 h	15.5	Mortality	Early instar	1,800 (800–5,200)	–	Sanders (1970)	LL/1, 5, 6
G. fasciatus	S	Nom	Technical	96 h	15.5	Mortality	Early instar	700 (190–8,200)	–	Sanders (1970)	LL/1, 5, 6

Gammarus lacustris	S	Nom	Technical	24 h	21.1	Mortality	2 months	700 (590–8,300)	–	Sanders (1969)	LL/1, 5, 6
G. lacustris	S	Nom	Technical	48 h	21.1	Mortality	2 months	380 (290–500)	–	Sanders (1969)	LL/1, 5, 6
G. lacustris	S	Nom	Technical	96 h	21.1	Mortality	2 months	160 (130–190)	–	Sanders (1969)	LL/1, 5, 6
Isochrysis galbana	S	Nom	Technical	3 days	20	Reduced oxygen evolution	Algal cells	10 (SE = 3)	–	Hollister and Walsh (1973)	LL/1, 2, 6
I. galbana Parke	S	Nom	Technical	90 min	20	Reduced oxygen evolution	Algal cells	10	–	Walsh (1972)	RL/1, 2, 8
I. galbana Parke	S	Nom	Technical	10 days	20	Growth inhibition	Algal cells	10	–	Walsh (1972)	RL/1, 2
Lemna gibba G3	S	Nom	98.0	7 days	25	Growth inhibition	NR	29 (27–31)	–	Okamura et al. (2003)	LR/6
Lemna minor	S	Nom	98.0	48 h	21	Reduced oxygen evolution	Plant fronds	–	LOEC = 5	Eullaffroy et al. (2007)	LL/1, 6, 7
L. minor 1769	S	Nom	98.0	7 days	25	Growth inhibition	NR	30 (28–31)	–	Okamura et al. (2003)	LR/6
L. minor	S	Nom	98.0	7 days	25	Growth inhibition	Plant fronds	25	LOEC = 5	Teisseire et al. (1999)	RL/1, 6
Lepomis macrochirus	S	Nom	Technical	96 h	12.7	Mortality	0.6–1.5 g	8,900 (8,200–9,600)	–	Macek et al. (1969)	LL/1, 5, 6
L. macrochirus	S	Nom	Technical	96 h	18.3	Mortality	0.6–1.5 g	7,600 (7,000–8,200)	–	Macek et al. (1969)	LL/1, 5, 6
L. macrochirus	S	Nom	Technical	96 h	23.8	Mortality	0.6–1.5 g	5,900 (5,300–6,500)	–	Macek et al. (1969)	LL/1, 5, 6

(continued)

Table 7 (continued)

Species	Test type	Meas/Nom	Chemical grade (%)	Duration	Temp (°C)	End point	Age/size	LC/EC$_{50}$ (μg/L) (95% CI)	MATC (μg/L)	Reference	Rating/reason
Lymnaea spp.	S	Nom	NR	96 h	NR	Mortality	Adult	15,300	–	Christian and Tate (1983)	LL/1, 3, 6
Monochrysis lutheri	S	Nom	Technical	3 days	20	Reduced oxygen evolution	Algal cells	18 (SE = 3)	–	Hollister and Walsh (1973)	LL/1, 2, 6
M. lutheri Droop	S	Nom	Technical	10 days	20.5	Growth inhibition	Algal cells	$EC_{100} = 0.02$	–	Ukeles (1962)	LL/1, 2, 6
M. lutheri Droop	S	Nom	Technical	10 days	20.5	Mortality	Early instar	2,500 (1,000–5,500)	–	Sanders (1970)	LL/1, 5, 6
Myriophyllum spicatum	S	Nom	>99	14 days	NR	Relative growth rate	Terminal lengths of shoots with 3 nodes	5	NOEC = 0.0005	Lambert et al. (2006)	LL/1, 5, 6
M. spicatum	S	Nom	>99	14 days	NR	Change in chlorophyll fluorescence ratio	Terminal lengths of shoots with 3 nodes	>5	NOEC = 5	Lambert et al. (2006)	LL/1, 5, 6
Navicula forcipata	S	Nom	99.0	96 h	20	Growth inhibition	Algal cells	27	–	Gatidou and Thomaidis (2007)	LL/2, 5
Navicula inserta	S	Nom	Technical	3 days	20	Reduced oxygen evolution	Algal cells	93(SE = 12)	–	Hollister and Walsh (1973)	LL/1, 2, 6
Neochloris sp.	S	Nom	Technical	3 days	20	Reduced oxygen evolution	Algal cells	28 (SE = 5)	–	Hollister and Walsh (1973)	LL/1, 2, 6
Nitzschia (Ind. 684)	S	Nom	Technical	3 days	20	Reduced oxygen evolution	Algal cells	169 (SE = 17)	–	Hollister and Walsh (1973)	LL/1, 2, 6

Nitzschia closterium	S	Nom	Technical	3 days	20	Reduced oxygen evolution	Algal cells	50 (SE = 6)	–	Hollister and Walsh (1973)	LL/1, 2, 6
Oscillatoria cf. chalybea	S	Nom	80.0	96 h	25	Growth inhibition	Algal cells	28	LOEC = 280	Schrader et al. (1998)	LR/1, 6
Oncorhynchus clarki (Salmo clarki)	S	Nom	95.0	96 h	10.0	Mortality	3.00 g	1,400 (1,100–1,900)	–	Johnson and Finley (1980)	LL/5, 6
Oncorhynchus mykiss (Salmo gairdneri)	S	Nom	95.0	96 h	13	Mortality	0.8 g	4,900 (4,100–5,900)	–	Johnson and Finley (1980)	LL/5, 6
O. mykiss (Salmo gairdneri)	S	Nom	80.0	96 h	13	Mortality	1.2 g	16,000 (11,300 –22,700)	–	Johnson and Finley (1980)	LL/5, 6
O. mykiss	S	Nom	95	7 days	10	Mortality	Juveniles, hatched <24 h ago	74,000 (29,000–3,681,000)	–	Okamura et al. (2002)	LR/1, 6
O. mykiss	S	Nom	95	14 days	10	Mortality	Juveniles, hatched <24 h ago	15,000 (11,000 –29,000)	–	Okamura et al. (2002)	LR/1, 6
O. mykiss	S	Nom	95	21 days	10	Mortality	Juveniles, hatched <24 h ago	5,900 (4,700–7,700)	–	Okamura et al. (2002)	LR/1, 6
O. mykiss	S	Nom	95	28 days	10	Mortality	Juveniles, hatched <24 h ago	230 (8.9–590)	–	Okamura et al. (2002)	LR/1, 6
Phaeodactylum tricornutum	S	Nom	Technical	3 days	20	Reduced oxygen evolution	Algal cells	10 (SE = 3)	–	Hollister and Walsh (1973)	LL/1, 2, 6
P. tricornutum Bohlin	S	Nom	Technical	90 min	20	Reduced oxygen evolution	Algal cells	10	–	Walsh (1972)	RL/1, 2, 8

(continued)

Table 7 (continued)

Species	Test type	Meas/Nom	Chemical grade (%)	Duration	Temp (°C)	End point	Age/size	LC/EC$_{50}$ (μg/L) (95% CI)	MATC (μg/L)	Reference	Rating/reason
P. tricornutum Bohlin	S	Nom	Technical	10 days	20	Growth inhibition	Algal cells	10	–	Walsh (1972)	RL/1, 2
P. tricornutum Bohlin	S	Nom	Technical	10 days	20.5	Growth inhibition	Algal cells	$EC_{21} = 0.4$	–	Ukeles (1962)	LL/1, 2, 6
Pimephales promelas	FT	Meas	98.6	96 h	24.3	Mortality	30 days	14,200 (13,400–15,000)	–	Call et al. (1983, 1987)	RL/1, 5
Platymonas sp.	S	Nom	Technical	3 days	20	Reduced oxygen evolution	Algal cells	7 (SE = 3)	–	Hollister and Walsh (1973)	LL/1, 2, 6
Porphyridium cruentum	S	Nom	Technical	3 days	20	Reduced oxygen evolution	Algal cells	24 (SE = 3)	–	Hollister and Walsh (1973)	LL/1, 2, 6
Protcoccus sp.	S	Nom	Technical	10 days	20.5	Growth inhibition	Algal cells	$EC_{48} = 0.02$	–	Ukeles (1962)	LL/1, 2, 6
Pseudokirchneriella subcapitata (*Selenastrum capricornutum*)	S	Nom	80.0	96 h	25	Growth inhibition	Algal cells	36.4	LOEC = 280	Schrader et al. (1998)	LR/1, 6
P. subcapitata (*S. capricorn utum*)	S	Nom	98.0	3 days	25	Growth inhibition	Algal cells	6.6 (5.9–7.2)	–	Okamura et al. (2003)	LL/5, 6
P. subcapitata (*S. capricorn utum*)	S	Nom	99.8	20 min	21.5	Change in chlorophyll fluorescence ratio	2–4-week-old algal cells	13.8 (9.3–20.4)	0.22	Podola and Melkonian (2005)	RL/1, 8

P. subcapitata (S. capricornutum)	S	Nom	98	120 h	24	Growth inhibition	Algal cells	22	NOEC = 10	Douglas and Handley (1988)	RL/6
P. subcapitata (S. capricornutum)	S	Nom	98	72 h	24	Growth inhibition	Algal cells	18	–	Douglas and Handley (1988)	RL 6
Pteronarcys californica	S	Nom	95.0	96 h	15	Mortality	Second year class	1,200 (900–1,700)	–	Johnson and Finley (1980)	LL/5, 6
P. californica	S	Nom	Technical	24 h	15.5	Mortality	30–35 mm	3,600 (2,800–4,700)	–	Sanders and Cope (1968)	LL/1, 5, 6
P. californica	S	Nom	Technical	48 h	15.5	Mortality	30–35 mm	2,800 (2,100–3,800)	–	Sanders and Cope (1968)	LL/1, 5, 6
P. californica	S	Nom	Technical	96 h	15.5	Mortality	30–35 mm	1,200 (870–1,700)	–	Sanders and Cope (1968)	LL 1, 5, 6
Raphidocelis subcapitata	S	Nom	50.0	96 h	25	Growth inhibition	Algal cells	0.7	–	Ma et al. (2006)	LL/3, 5, 6
Salvelinus namaycush	S	Nom	95.0	96 h	10	Mortality	1.5 g	2,700 (2,400–3,000)	–	Johnson and Finley (1980)	LL/5, 6
Scenedesmus obliquus	S	Nom	50.0	96 h	25	Growth inhibition	Algal cells	4.09	–	Ma (2002)	LL/1, 3, 6
S. obliquus	S	Nom	98.0	1 min	22	Change in chlorophyll fluorescence ratio	Algal cells	1[a]	–	Eullaffroy and Vernet (2003)	LL/1, 4, 6, 8

(continued)

Table 7 (continued)

Species	Test type	Meas/ Nom	Chemical grade (%)	Duration	Temp (°C)	End point	Age/size	LC/EC$_{50}$ (μg/L) (95% CI)	MATC (μg/L)	Reference	Rating/ reason
Scenedesmus quadricauda	S	Nom	50.0	96 h	25	Growth inhibition	Algal cells	2.7	–	Ma (2003)	LL/1, 3, 6
Scenedesmus subspicatus	S	Nom	Technical	24 h	20	Growth inhibition	Algal cells, 3-day old	NR	NOEC = 4	Schafer et al. (1994)	LR/5, 6
S. subspicatus	S	Nom	Technical	72 h	20	Growth inhibition	Algal cells, 3-day old	36	NOEC = 10	Schafer et al. (1994)	LR/5, 6
Scherffelia dubia	S	Nom	99.8	20 min	21.5	Change in chlorophyll fluorescence ratio	2–4-week-old algal cells	3.9 (2.5–6.2)	0.22	Podola and Melkonian (2005)	RL/1, 8
Simocephalus serrulatus	S	Nom	95.0	48 h	15	Mortality	First instar	2,000 (1,400–2,800)	–	Johnson and Finley (1980)	LL/5, 6
Staurodesmus convergens	S	Nom	99.8	20 min	21.5	Change in chlorophyll fluorescence ratio	2–4-week-old algal cells	4.1 (2.5–6.9)	0.22	Podola and Melkonian (2005)	RL/1, 5, 8
Stauroneis amphoroides	S	Nom	Technical	3 days	20	Reduced oxygen evolution	Algal cells	31 (SE = 2)	–	Hollister and Walsh (1973)	LL/1, 2, 6
Synechocystis sp.	S	Nom	99.8	20 min	21.5	Change in chlorophyll fluorescence ratio	2–4-week-old algal cells	7.6 (5.5–10.5)	0.22	Podola and Melkonian (2005)	RL/1, 5, 8
Tetraselmis elegans	S	Nom	99.8	20 min	21.5	Change in chlorophyll fluorescence ratio	2–4-week-old algal cells	3.0 (2.3–3.8)	0.22	Podola and Melkonian (2005)	RL/1, 8

Thalassiosira fluviatilis	S	Nom	Technical	3 days	20	Reduced oxygen evolution	Algal cells	95 (SE = 10)	–	Hollister and Walsh (1973)	LL/1, 2, 6
Ulothrix fimbriata	S	Nom	95.0	7 days	25	Growth inhibition	Algal cells	540	–	Maule and Wright (1984)	LR/1, 6

S Static, *SR* static renewal, *FT* flow through, *NR* not reported, *95% CI* 95% confidence interval, *SE* standard error

Reasons for ratings

1. Not a standard method
2. Saltwater
3. Low chemical purity or purity not reported
4. Toxicity value not calculable
5. Control not described and/or response not reported
6. Low reliability score
7. End point not linked to growth, reproduction, or survival
8. Inappropriate test duration

[a] Value reported as toxicity threshold, which is conceptually very similar to an MATC, but calculated differently than an MATC or an EC_x

[b] Growth inhibition of roots is not a standard end point

including the rationale for the scores and ratings were created for each study, all of which are included in the Supporting Material (http://extras.springer.com/).

Because diuron is a herbicide, many of the single-species studies were plant toxicity tests. Plant data are more difficult to interpret than animal data because a variety of end points may be used, but the significance of each one is not clear. According to the UCDM, all plant studies were considered as chronic because the typical end points of growth or reproduction are inherently chronic. Only end points of growth or reproduction (measured by biomass) and tests lasting at least 24 h had the potential to be rated highly, and to be used for criteria calculation, which is in accordance with standard methods (ASTM 2007a, 2007b, USEPA 1996). The four main end points identified in plant toxicity tests are described below, including whether the end point is clearly linked to survival, growth, or reproduction.

2.1 Growth Inhibition

All of these end points are evaluated relative to a control growth measurement. Depending on the plant, the endpoint measurement may have been assessed by direct cell counts with a hemacytometer, cell counts with a spectrophotometer, cell counts with an electronic particle counter, chlorophyll concentration measured by absorbance, turbidity measured by absorbance, or number of fronds (*Lemna* spp.). In all cases, growth of exposed samples was compared statistically to controls.

2.2 Relative Growth Rate

The biomass of macrophytes was measured before and after exposure to calculate a growth rate as (final mass–initial mass)/initial mass $\times$ 100. This end point is very similar to growth inhibition, except that it is expressed as a positive effect while growth inhibition is expressed as a negative effect. In all cases, the growth rate of exposed samples was compared statistically to controls.

2.3 Change in Chlorophyll Fluorescence Ratio

Chlorophyll fluorescence was measured at a maximal fluorescence and either a variable or steady-state fluorescence and a ratio were computed. An increase in the ratio indicates a disruption of photosystem II (PSII), which may lead to a decrease in carbohydrate production and thus decreased growth. With this end point, one measures physiological stress in plants (Lambert et al. 2006). This ratio is a valid measurement that is related to algal growth according to ASTM Standard Method E1218-04 (ASTM 2004), but is described as being less definitive than measuring chlorophyll *a* content, and is therefore not a preferred end point if one more directly related to growth is available.

2.4 Reduced Oxygen Evolution

Plants evolve oxygen during photosynthesis, and reduced photosynthesis has been shown by Walsh (1972) to correlate well with the concentrations that inhibit growth, but it is not clear that this end point is a good predictor of growth inhibition across all plant species. The value for this end point is always calculated as being relative to controls.

To ensure that the derived criteria are protective of ecosystems and used all available data, all multispecies mesocosm, microcosm, and ecosystem (field and laboratory) studies that were rated as being acceptable and reliable (R) or less reliable (L) were compared to the criteria. Studies on the effects of diuron on mallard ducks were rated for reliability using the terrestrial wildlife evaluation table. Mallard studies that were rated as being reliable (R) or less reliable (L) were used to evaluate the bioaccumulation of diuron.

3 Data Reduction

The data reduction procedure is described by Palumbo et al. (2012). Multiple toxicity values for diuron for the same species were reduced down to a species mean acute value (SMAV) or a species mean chronic value (SMCV). Acceptable (RR) data were excluded from the final data sets that were employed for criteria calculations for the following reasons: more appropriate exposure durations were available, flow-through tests are preferred over static tests, a test with a more sensitive life stage of the same species was available, and tests with more sensitive end points were available. Excluded data are given in Table 6. The final acute data set contains three animal SMAVs (Table 3), the final chronic plant data set contains three SMCVs (Table 4), and the final chronic animal data set contains ten SMCVs (Table 5).

4 Acute Criterion Calculation

Although plants are more sensitive to diuron, the acute criterion was calculated from acute animal toxicity data because plant toxicity tests are considered as being chronic. Three SMAVs from two different taxa were available: planktonic crustaceans (*Daphnia magna* and *Daphnia pulex*) and a benthic invertebrate (*Hyalella azteca*). Because there were so few data, the acute criterion was not calculated using a species sensitivity distribution (SSD). At least five data values are required to fit an SSD to a data set, and the data must fulfill five different taxa requirements (planktonic crustacean, benthic invertebrate, fish from the family Salmonidae, warm water fish, and insect). Instead, the acute criterion was calculated using the assessment factor (AF) procedure (TenBrook et al. 2010). The AFs in the UCDM were derived by randomly sampling 12 organic pesticide data sets to give estimates of the median fifth percentile of a distribution (TenBrook et al. 2010). AFs are recognized as a conservative approach for dealing with uncertainty in assessing risks posed by chemicals and are widely used in other methods for deriving criteria.

The acute criterion was calculated by dividing the lowest SMAV (12 mg/L for *D. magna*) from the acceptable (RR) data set by an AF. The magnitude of the AF was determined by the number of taxa available in the data set. The acute data set fulfilled two of the five taxa requirements, corresponding to an AF of 36 (TenBrook et al. 2010). The acute value calculated using the AF represents an estimate of the median fifth percentile of the SSD, which is the recommended acute value. The recommended acute value is divided by a factor of 2 to calculate the acute criterion. Because the toxicity datum used to calculate the criterion was presented in only two significant figures, the criterion is rounded to two significant figures.

$$\begin{aligned}\text{Acute value} &= \frac{\text{LowestSMAV}}{\text{Assessment factor}},\\ &= 0.33\ \text{mg/L}.\end{aligned} \tag{1}$$

$$\begin{aligned}\text{Acute criterion} &= \frac{\text{Acutevalue}}{2},\\ &= 0.17\ \text{mg/L}\ (170\ \ \mu\text{g/L}).\end{aligned} \tag{2}$$

5 Chronic Criterion Calculation

The chronic data demonstrate that plants are more sensitive to diuron than animals. Because diuron is a herbicide and the data demonstrates that plants are the most sensitive taxon, only plant data were used to derive the chronic criterion. The chronic criterion is likely to also be protective of animals because they are less sensitive to diuron. Four acceptable maximum acceptable toxicant concentrations (MATCs) and five acceptable EC_{50}s were available for vascular plants or alga. MATCs are recommended for derivation of the chronic criterion because they approximate a no-effect concentration (unlike EC_{50}s). EC_x toxicity values are not recommended for chronic criteria derivation unless there is data for the relevant species indicating what level of x corresponds to a no-effect level, which was not available for the diuron data set. Since there were too few MATCs to fit a distribution to the data, the chronic criterion was derived by setting the chronic criterion equal to the lowest

NOEC from an important alga or vascular aquatic plant species that has measured concentrations and a biologically relevant end point (TenBrook et al. 2010). In this scheme, the NOEC of 1.3 μg/L for the green algae *Pseudokirchneriella subcapitata* (formerly *Selenastrum capricornutum*) serves as the chronic criterion.

6 Water Quality Effects and Bioavailability

Temperature and pH do not appear to have a significant effect on the toxicity of diuron, as it is only a very weak base and no such effects have been documented in the literature. Because diuron has a moderate octanol–water partition coefficient (log $K_{ow} = 2.78$), decreased bioavailability due to surface sorption is possible. Knauer et al. (2007) demonstrated that the addition of black carbon (BC) in its native form to water only slightly decreased the toxicity of diuron to the freshwater green algae *P. subcapitata* (formerly *S. capricornutum*). BC is ubiquitous in the environment because it is a product of incomplete combustion and can act as a supersorbent for some organic contaminants as a result of its large surface area, but it represents only a small fraction of total organic carbon, which is usually responsible for the majority of sorption to solids. Studies in which the sorption of diuron to dissolved organic carbon and clays were investigated are not currently available in the literature, but sorption to these materials is also likely to inhibit bioavailability in a similar manner as sorption to BC. Because there is little information regarding which phases of diuron (freely dissolved, sorbed to dissolved organic carbon, or sorbed to suspended solids) are bioavailable, it is recommended that criteria compliance is based on whole water concentrations.

7 Chemical Mixtures

Diuron is a PSII inhibitor, as are all phenylurea herbicides. Other widely used herbicides, such as the triazines, are also PSII inhibitors, but have different binding sites than the phenylurea herbicides. The concentration addition model is recommended because it has been tested and shown to successfully predict the toxicity of compound mixtures that possess the same mode of action (Mount 2003). It has been confirmed in several studies that the toxicity of a mixture of PSII-inhibitor herbicides, including diuron, can be predicted by the concentration addition method (Arrhenius et al. 2004; Backhaus et al. 2004; Knauert et al. 2008). When diuron is detected with other PSII-inhibitor herbicides, the toxicity of the mixture should be predicted by the concentration addition model and used to determine criteria compliance. If numeric water quality criteria are not available for other PSII-inhibitor herbicides, the model cannot be used and diuron should be considered alone.

The toxicity of diuron in mixtures with other chemicals that work by different modes of action has been reported (e.g., Hernando et al. 2003; Walker 1965), but interaction coefficients for multiple species have not been calculated. Therefore, nonadditive mixture toxicity cannot yet be incorporated into criteria compliance. Lydy and Austin (2005) demonstrated a nonadditive form of toxicity when mixtures of diuron and organophosphate insecticides were tested; these authors found that some acted as synergists with diuron. Teisseire et al. (1999) examined the phytotoxicity of the herbicide combined with two fungicides (copper and folpet) on duckweed (*Lemna minor*) because these pesticides are often used in combination in vineyards. They found that growth inhibition from the combination of diuron and copper depended on the concentrations of both chemicals used, whereas it only depended on the herbicide's concentration when combined with folpet. Diuron is widely used as an antifouling biocide in paint for ship hulls and is often used in combination with other antifouling agents. Several articles were found in which researchers studied the toxicity of mixtures of diuron or diuron metabolites and other antifouling agents, including Irgarol (cybutryne), Sea nine 211 (4, 5-dichloro-2-*n*-octyl-3(2H)-isothiazolone), copper, chlorothalonil, copper pyrithione, zinc pyrithione, and tri-*n*-butyltin (Chesworth et al. 2004; Fernandez-Alba et al. 2002; Gatidou and Thomaidis 2007; Koutsaftis and Aoyama 2007; Manzo et al. 2008; Molander et al. 1992). Resulting toxicities were synergistic, additive, or antagonistic for different mixtures, and were sometimes dependent on concentration ratios and how many compounds were in the mixture.

8 Sensitive Species

The derived criteria were compared to the most sensitive toxicity values in both the acceptable (RR) and supplemental (RL, LR, LL) data sets to ensure that these species are adequately protected. The lowest acute value in the data sets is 160 μg/L for the amphipod *Gammarus lacustris* (Sanders 1969), which is below the derived acute criterion of 170 μg/L. This study was rated LL because the control response was not reported, many other study details were not documented, and the test concentrations were not measured. Additionally, data for another amphipod, *Gammarus fasciatus*, is the next lowest acute value in the data set (700 μg/L), indicating that *Gammarus* species are particularly sensitive to diuron. Because the *G. lacustris* toxicity value is based on nominal, instead of measured, concentrations, the acute criterion was not adjusted downward. If measured data that is highly rated becomes available for *Gammarus* species in the future, it should be examined to determine if the acute criterion is protective of this sensitive genus.

Although there are several supplemental chronic data values that are below the derived chronic criterion (1.3 μg/L), the criterion was not adjusted because the lower toxicity values were lacking at least one of the following critical parameters: (1) the use of an end point that directly related to survival, growth, or reproduction; (2) the use of an exposure duration of $\geq$24 h (ASTM 2007a, 2007b; USEPA 1996);

(3) proper design of hypothesis tests and reporting of parameters used to evaluate the reasonableness of the resulting toxicity values; (4) the use of diuron $\geq$80% purity; and (5) the use of freshwater species. These studies are discussed in detail below.

The lowest measured chronic value in the data sets is an EC_{50} of 0.00026 µg/L for the rooted macrophyte *Apium nodiflorum*—for a nonstandard end point of root growth (Lambert et al. 2006). This value was calculated by extrapolation, not interpolation, is lower than the NOEC reported for this test, and is below the lowest concentration tested; thus, it was not used for criterion adjustment. There are several other NOECs reported in this study for an appropriate end point (relative growth rate) that are below the proposed chronic criterion (0.0005–0.05 µg/L), but it was not possible to evaluate the reasonableness of these NOECs because the control responses were not reported, the *p*-value selected was not reported, and a minimum significant difference was not calculated.

Podola and Melkonian (2005) report NOEC and LOEC values of 0.1 and 0.5 µg/L, respectively, for nine different algae. These values are below the proposed criteria, but this study used a less preferred end point, change in chlorophyll fluorescence, and a nonstandard exposure duration of 20 min. The authors proposed the use of a biosensor to detect and identify herbicides in the environment, and do not discuss the link between the effects they quantify and survival, growth, or reproduction of the algal strains. Similarly, Eullaffroy and Vernet (2003) reported a toxicity threshold of 1 µg/L for green algae, which is slightly below the chronic criterion. The exposure duration was only 1 min, and its purpose was to rapidly detect herbicides in the environment. This study did not follow a standard method, used extremely short exposure durations, and did not include an acceptable toxicity value (e.g., NOEC, LOEC, MATC, or EC_x). Values from these studies cannot be directly related to survival, growth, or reproduction, and probably only demonstrate exposure to diuron, not adverse effects. Therefore, the chronic criterion was not adjusted downward based on these data.

Ma et al. (2001) and Ma (2002) performed studies that contained the same data for the alga *Chlorella pyrenoidosa*, an EC_{50} equal to the derived criterion. These studies used diuron with a purity of 50% and did not report a control response. In another study by Ma et al. (2006), an EC_{50} below the derived criterion (0.7 µg/L) was reported, but also used diuron of 50% purity. The low-purity compound used in these tests precludes the use of them for criterion adjustment. One study that used saltwater organisms (Ukeles 1962) reported toxicity values below the derived chronic criterion (0.02 and 0.4 µg/L), but such organisms are suspected to have different sensitivities than freshwater species; therefore, they are not used to derive or adjust freshwater criteria.

9 Ecosystem-Level Studies

The chronic criterion was compared to multispecies studies to ensure that the results from single-species studies are protective of multispecies systems. Ten mesocosm, microcosm, or ecosystem (field and laboratory) studies were identified (Table 10), which were almost all indoor or laboratory studies mimicking small river or pond natural environments and in which microbial, phytoplanktonic, or bacterial communities were examined. An initial drop in phytoplankton biomass was noted in most of these studies, which led to a decrease in dissolved oxygen from the decay of the phytoplankton.

Planktonic communities have displayed varying degrees of response to diuron, depending on, among other things, the concentrations applied. Hartgers et al. (1998) set up microcosms containing phyto-, peri-, bacterio-, and zoo-plankton and monitored them for a 28-day exposure to a mixture of diuron, atrazine, and metolachlor, followed by a 28-day recovery period. An NOEC for the mixture based on phytoplankton was determined to be 1.5 μg/L diuron; thus, the criterion of 1.3 μg/L would likely be protective of phytoplankton based solely on diuron. Flum and Shannon (1987) reported a 96-h EC_{50} of 2,205 μg/L (1,630–3,075 μg/L 95% CI) for an artificial microecosystem containing zooplankton, amphipods, ostracods, unicellular and filamentous algae, protozoans, and microbes, which is much higher than the derived chronic criterion. The EC_{50} was based on monitoring the redox potential, pH, and dissolved oxygen as a measure of toxicity.

Planktonic and algal communities exposed to diuron have been studied in regard to the aquaculture industry because some algae give fish an "off" flavor, yet plankton is necessary for healthy ponds. Zimba et al. (2002) assessed the effect of 9 weeks of diuron application (10 μg/L) on catfish pond ecology. The only significant effect from the exposure was a change in the phytoplankton composition; its biomass was not altered. Perschbacher and Ludwig (2004) also studied plankton communities in outdoor pool mesocosms simulating aquaculture ponds. Three diuron concentrations were tested and monitored for 4-weeks post application. Diuron depressed primary production and biomass of phytoplankton for at least 4-weeks post application, which in turn caused a decrease in dissolved oxygen to levels that are potentially lethal to fish. The concentrations were not measured, and were reported as field rate (1.4 kg a.i./ha), 1/10 field rate, and 1/100 field rate of Direx without adjuvants.

Tlili et al. (2008) studied biofilm communities in a small river with chronic exposure to 1 μg/L diuron, as well as 3-h pulses of 7 or 14 μg/L diuron with and without prior exposure. The results indicate that photosynthesis was never significantly inhibited by any of the treatments, but the pulses did alter the community structure of the microalgae. The pulses affected the eukaryotic community structure in microcosms that did not have prior chronic diuron exposure, but had no significant impact on those that did have prior exposure. Dorigo et al. (2007) assessed prokaryotic and eukaryotic communities and microalgae exposed to vineyard runoff water in a small stream containing diuron concentrations of 0.09 and 0.43 μg/L. The diuron tolerance in these communities increased in the downstream direction and the pristine control site had the lowest tolerance, following the concept that contaminant exposure increases the tolerance of biofilms either by adaptation or species changes. The end points in these studies are not clearly linked to survival, growth, and reproduction and do not exhibit a clear dose–response relationship, so it is not clear if diuron exposure at these levels impacted the diversity of species in biofilm communities. Community restructuring may have long-term effects on an ecosystem; however, the studies available only provide preliminary data on this subject. The authors of two other studies also reported adverse effects on microbes from diuron exposure (Pesce et al. 2006; Sumpono et al. 2003), but the concentrations tested were well-above the derived criteria and do not provide information regarding protection at levels near the criterion.

The literature shows that herbicides in aquatic ecosystems may have detrimental effects on the bottom trophic levels of the food chain, which may indirectly impact species up the food chain via changes in water quality or decreased food supply. However, many of these studies only tested a single concentration, and no dose–response relationship can be inferred and no-effect concentrations are not available. Considering the available studies, it appears that the derived acute and chronic criteria could be protective of these types of negative effects because most studies used much higher exposure concentrations. The only studies that reported effects at concentrations lower than the derived chronic criterion examined biofilm community restructuring, and provided preliminary data that cannot be incorporated into criteria derivation until more in-depth studies are available.

10 Threatened and Endangered Species

Threatened and endangered species (TES) may be more sensitive than standard test species, and their protection is considered by comparing toxicity values for TES to the derived criteria. Several listed animal species are represented in the data set (CDFG 2010a, 2010b; USFWS 2010). There is an RR study for *Rana aurora*, which has a related subspecies that is endangered (California red-legged frog, *R. a. draytonii*). The *R. aurora* 14-day LC_{50} is 22.2 mg/L, which is well above the acute criterion of 0.17 mg/L. The supplemental data set includes acute toxicity values for the listed salmonids *Oncorhynchus mykiss* and *Oncorhynchus clarki* (listed subspecies is *Oncorhynchus clarki henshawi*). There are two 96-h LC_{50}s for *O. mykiss* of 4.9 (4.1–5.9) mg/L and 16 (11.3–22.7) mg/L, and an LC_{50} of 1.4 (1.1–1.9) mg/L for cutthroat trout (*O. clarki*), which are both well above the acute criterion of 0.17 mg/L.

The USEPA interspecies correlation estimation (Web-ICE v. 3.1; Raimondo et al. 2010) software was used to estimate toxicity values for the listed animals represented in the acute data set by members of the same family or genus. The estimated toxicity values (Table 8) range from 0.729 to 4.491 mg/L for various salmonids.

No plant studies used in the criteria derivation were performed on state or federal endangered, threatened, or rare species. Plants are particularly sensitive to diuron because it is a herbicide, but there are no aquatic plants listed as state or federal endangered, threatened, or rare species; so they could not be considered in this section.

Table 8 Threatened, endangered, or rare species predicted values by Web-ICE (v. 3.1; Raimondo et al. 2010)

Surrogate		Predicted	
Species	LC_{50} (mg/L)	Species	LC_{50} (95% confidence interval) (mg/L)
Rainbow trout (*Oncorhynchus mykiss*)	4.9	*Oncorhynchus aguabonita whitei*	4.491 (3.613–5.581)
			4.491 (3.613–5.581)
		Oncorhynchus gilae apache	4.491 (3.613–5.581)
		Oncorhynchus gilae	4.491 (3.613–5.581)
		Oncorhynchus nerka	5.983 (3.225–11.097)
		Oncorhynchus tshawytscha	8.086 (6.104–4.016)
		Oncorhynchus kisutch	4.758 (3.545–6.387)
		Oncorhynchus clarki henshawi	
Cutthroat trout (*O. clarki*)	1.4	*Oncorhynchus clarkii henshawi*	1.206 (0.967–1.504)
			1.206 (0.967–1.504)
		Oncorhynchus clarkii seleniris	1.206 (0.967–1.504)
			0.729 (0.290–1.832)
		Oncorhynchus clarkii stomias	0.729 (0.290–1.832)
			1.673 (1.156–2.421)
		O. gilae apache	1.206 (0.967–1.504)
		O. gilae	1.206 (0.967–1.504)
		O. kisutch	
		O. nerka	
		O. tshawytscha	

11 Bioaccumulation and Partitioning to Air and Sediment

Diuron has a log K_{ow} of 2.78 (Sangster Research Laboratories 2008), and a molecular weight of 233.1, which indicates a low bioaccumulative potential. There is a USEPA pesticide tolerance established for farm-raised freshwater finfish tissue of 2.0 mg/kg (USEPA 2007), but there are no FDA food tolerances for diuron (USFDA 2000). The bioconcentration of diuron has been measured in various species (Table 9) and these bioconcentration factors (BCFs) indicate that it has a low potential for bioaccumulation in the environment. Because diuron has a low potential to bioaccumulate and low toxicity to mallard ducks (lowest dietary $LC_{50} = 1{,}730$ mg/kg feed; USEPA 2003), the protection of terrestrial wildlife from bioaccumulation was not assessed further. Because diuron has a low vapor pressure and a moderate log K_{ow}, it is also not likely to partition to the air or sediment, and currently there were no state or federal air quality or sediment quality standards identified for diuron (CARB 2008; CDWR 1995; NOAA 1999).

Table 9 Bioconcentration factors (BCFs) for diuron

Species	BCF	Exposure	Reference
Gambusia affinis	290	S	Isensee (1976)
Physa sp.	40	S	Isensee (1976)
Daphnia magna	260	S	Isensee (1976)
Oedogonium cardiacum	90	S	Isensee (1976)
Pimephales promelas	2.00	FT	Call et al. (1983, 1987)

FT flow through, *S* static
Values are on a wet weight basis and are not lipid normalized

Table 10 Acceptable multispecies field, semifield, laboratory, microcosm, mesocosm studies

Reference	Habitat	Rating
Devilla et al. (2005)	Laboratory model ecosystem	L
Dorigo et al. (2007)	Lotic outdoor stream	L
Flum and Shannon (1987)	Laboratory microcosm	L
Hartgers et al. (1998)	Laboratory microcosm	R
Molander and Blanck (1992)	Laboratory microcosm	L
Perschbacher and Ludwig (2004)	Outdoor pond	L
Pesce et al. (2006)	Laboratory microcosm	L
Sumpono et al. (2003)	Indoor pond	R
Tlili et al. (2008)	Laboratory microcosm	R
Zimba et al. (2002)	Outdoor pond	L

R reliable, *L* less reliable

12 Assumptions, Limitations, and Uncertainties

Environmental managers have the discretion to choose how to use water quality criteria, as such, they should be aware of the assumptions, limitations, and uncertainties involved in the calculations, and the accuracy and confidence in criteria. The UCDM (TenBrook et al. 2010) identifies these points for the various recommended procedures, and this section summarizes any specific data limitations that affected the procedure used to determine the final diuron criteria.

One major limitation was the lack of highly rated acute toxicity data for diuron, which prevented the use of an SSD for acute criterion derivation. Only two of the five taxa required for use of an SSD were available; the three missing taxa were a warm water fish, a fish from the family Salmonidae, and an insect. Because of this lack of data, an AF was used to calculate the acute criterion. Uncertainty cannot be quantified using the AF procedure, as it is based on only one toxicity value. There were no highly rated amphipod data available, which is an important data gap, as this taxon appears to be the most sensitive animal taxa.

The most important limitation is the lack of acceptable plant data because plants are much more sensitive to diuron than animals. Plant and algal data can be difficult to interpret and do not use consistent end points. The chronic data set contained five EC_{50}s and four MATCs, which are the preferred toxicity values for chronic tests.

The methodology requires that MATCs are used to derive chronic criteria by the SSD procedure, unless studies are available with EC_x values that show what level of x is appropriate to represent a no-effect level. Thus, the chronic criterion was calculated as the lowest NOEC in the data set. In this approach, the chronic criterion was derived with the absolute minimum amount of data, and uncertainty cannot be quantified because it is based on only one toxicity value.

Other limitations include the lack of information about diuron and mixture toxicity and ecosystem-level effects. There is evidence that diuron exhibits synergism with some other chemicals, including organophosphate pesticides, but there is a lack of multispecies interaction coefficients available to incorporate the presence of chemical mixtures into criteria compliance. Biofilms displayed sublethal effects to low-level diuron exposures, but these effects need to be further investigated to determine if the exposures are linked to survival, growth, or reproduction of organisms in biofilms. Another issue to consider is the averaging periods of the acute and chronic criteria. The chronic 4-day averaging period should be protective based on available data. However, the acute criterion is very high when compared to plant data, and it may allow for a pulse that could kill off a large amount of algae, resulting in increased biological demand and potential fish kills due to low dissolved oxygen, as discussed in Sect. 9. Clear data on the timing and concentrations that could cause this effect are not currently available, but should be considered when more data is available.

13 Comparison to Existing Criteria

The European Union has derived an environmental quality standard for diuron of 20 μg/L as a maximum allowable concentration and 2 μg/L as the annual average (Killeen 1997), which are analogous to the acute and chronic criterion, respectively. The maximum allowable concentration is lower than the UCDM acute criterion of 170 μg/L, and the annual average is very similar to the UCDM chronic criterion of 1.3 μg/L. These criteria were derived using safety factors, which are analogous to assessment factors. A safety factor of 10 was applied to the lowest credible lethal concentration, which was an LC_{50} of 160 μg/L for *G. fasciatus*, to calculate the maximum allowable concentration. A safety factor of 100 was applied to this datum to calculate the annual average. The authors noted that while algae demonstrated higher sensitivity to diuron, the effects on algae were algistatic, not algicidal, and that based on the algal data the environmental quality standards derived from the animal data are sufficiently protective of these species.

The Netherlands has derived a maximum permissible concentration (MPC) for diuron of 0.43 μg/L (Crommentuijn et al. 2000), which is analogous to a UCDM chronic criterion. This MPC was derived using a statistical extrapolation on the combined freshwater and marine data set, which included data for algae, crustaceans, insects, plants, and fish (Crommentuijn et al. 1997). The lowest reported NOEC was 0.056 μg/L for *Scenedesmus subspicatus*, which is more sensitive than any data in the acceptable UCDM data set.

14 Comparison to the USEPA 1985 Method

Water quality criteria for diuron were also calculated by using the USEPA (1985) method, which requires a total of eight taxa to use an SSD—three additional taxa beyond the five required by the UCDM. Only two of the eight total acute taxa requirements were fulfilled, a planktonic crustacean (*D. magna* or *D. pulex*) and a benthic invertebrate (*H. azteca*). Because of this lack of data, no diuron acute criterion could be calculated according to the USEPA (1985) methodology.

According to the USEPA (1985) methodology, the chronic criterion is equal to the lowest of the Final Chronic Value, the Final Plant Value, and the Final Residue Value. To calculate the Final Chronic Value, animal data is used and the same taxa requirements must be met as in the calculation of the acute criterion. Seven of the eight taxa requirements are available in the RR chronic animal data set (Table 5). The missing taxon is a fish from the family Salmonidae; the seven available taxa are as follows: (1) planktonic crustacean (*D. pulex*), (2) benthic invertebrate (*H. azteca*), (3) insect (*Chironomus tentans*), (4) warm water fish (*Pimephales promelas*), (5) a third family in the phylum Chordata (*Pseudacris regilla*, *R. aurora*, *Rana catesbeiana*, or *Xenopus laevis*), (6) a family in a phylum other than Arthropoda or Chordata (*Physa* sp.), and (7) a family in any order of insect or any phylum not already represented (*Lumbriculus variegatus*).

The California Department of Fish and Game has derived criteria using the USEPA (1985) SSD method with fewer than the eight required families, using professional judgment to determine that species in the missing categories were relatively insensitive and their addition would not lower the criteria (Menconi and Beckman 1996; Siepmann and Jones 1998). It is not clear that a fish from the family Salmonidae would be relatively insensitive to diuron because the lowest animal chronic toxicity value is for a fish (*P. promelas*). As an example, the data in Table 5 were used to calculate genus mean chronic values from the given SMCVs, and the log-triangular distribution was employed to yield a fifth percentile estimate.

$$\begin{aligned} \text{Final Chronic Value} &= \text{Fifth percentile estimate,} \\ &= 23\ \mu\text{g/L.} \end{aligned}$$

The Final Plant Value is calculated as the lowest result from a 96-h test conducted with an important plant species, in which the concentrations of test material were measured and the end point was biologically important. None of the plant toxicity values in the RR data set (Table 4) are for a 96-h test, and two use measured concentrations. The closest test that fits this description is the 120-h NOEC of 1.3 μg/L reported for *P. subcapitata* (Blasberg et al. 1991). This test has an exposure duration that is 24 h longer than the specified duration.

$$\begin{aligned} \text{Final Plant Value} &= \text{Lowest result from a plant test,} \\ &= 1.3\ \mu\text{g/L.} \end{aligned}$$

The Final Residue Value is calculated by dividing the maximum permissible tissue concentration by an appropriate BCF or bioaccumulation factor (BAF). A maximum allowable tissue concentration is either (a) an FDA action level for fish oil or for the edible portion of fish or shellfish or (b) a maximum acceptable dietary intake based on observations on survival, growth, or reproduction in a chronic wildlife feeding study or long-term wildlife field study. While no FDA action level exists for fish tissue, there is an EPA pesticide tolerance for farm-raised freshwater finfish tissue of 2.0 mg/kg (USEPA 2007). There is no relevant study that meets the requirement of part (b) above. A BCF of 2.0 for *P. promelas* (Table 9) was used to calculate the Final Residue Value.

$$\begin{aligned}\text{Final Residue Value} &= \frac{\text{Maximum permissible tissue concentration}}{\text{BCF}},\\ &= 1\ \text{mg/L}\ (1{,}000\ \mu\text{g/L}).\end{aligned}$$

The Final Plant Value is lower than both the Final Chronic Value and the Final Residue Value; therefore, the chronic criterion by the USEPA (1985) methodology would be 1.3 μg/L, and the example USEPA chronic criterion is equivalent to the UCDM chronic criterion.

15 Summary and Final Criteria Statement

Acute and chronic water quality criteria for the protection of aquatic life were derived for diuron using the UCDM. The acute criterion is based only on acute animal data and was derived using an assessment factor because there were insufficient data to use a SSD while the chronic criterion was derived using only plant data, which are more sensitive to diuron. The lowest NOEC of a highly rated plant study was used as the criterion because there were insufficient data for use of an SSD for criterion calculation. Plant toxicity data are essential when considering diuron usage and regulations because plants and algae are the most sensitive taxa; however, plant data are difficult to interpret. The criteria should be updated whenever relevant and reliable new data become available.

Aquatic life in the Sacramento River and San Joaquin River basins should not be affected unacceptably if the 4-day average concentration of diuron does not exceed 1.3 μg/L (1,300 ng/L) more than once every 3 years on the average and if the 1-h average concentration does not exceed 170 μg/L more than once every 3 years on the average. Mixtures of diuron and other PSII-inhibitor herbicides should be considered to be additive (see Sect. 7).

Acknowledgments We thank the following reviewers: D. McClure (CRWQCB-CVR), J. Grover (CRWQCB-CVR), S. McMillan (CDFG), J. P. Knezovich (Lawrence Livermore National Laboratory), and X. Deng (CDPR). This project was funded through a contract with the Central Valley Regional Water Quality Control Board of California. Funding for this project was provided by the

California Regional Water Quality Control Board, Central Valley Region (CRWQCB-CVR). The contents of this document do not necessarily reflect the views and policies of the CRWQCB-CVR, nor does mention of trade names or commercial products constitute endorsement or recommendation for use.

References

Arrhenius A, Gronvall F, Scholze M, Backhaus T, Blanck H (2004) Predictability of the mixture toxicity of 12 similarly acting congeneric inhibitors of photosystem II in marine periphyton and epipsammon communities. Aquat Toxicol 68:351–367.

ASTM (2004) Standard Guide for Conducting Static Toxicity Tests with Microalgae. In: ASTM E1218 (Environmental Toxicology Standards). American Society for Testing and Materials.

ASTM (2007a) Standard Guide for Conducting Static Toxicity Tests with Microalgae. Designation: E 1218–07. American Society for Testing and Materials.

ASTM (2007b) Standard Practice for Algal Growth Potential with *Pseudokirchneriella subcapitata*. Designation: D 3978–07. American Society for Testing and Materials.

Backhaus T, Faust M, Scholze M, Gramatica P, Vighi M, Grimme LH (2004) Joint algal toxicity of phenylurea herbicides is equally predictable by concentration addition and independent action. Environ Toxicol Chem 23:258–264.

Baer KN (1991) Static, Acute 48-hour EC50 of DPX-14740-165 (Karmex DF) to *Daphnia magna*. Haskell laboratory report No. 508-91. Unpublished study prepared by E. I. du Pont de Nemours and Company, Newark, DE, submitted to the U. S. Environmental Protection Agency. EPA MRID 42046003.

Blasberg J, Hicks SL, Bucksaath J (1991) Acute Toxicity of Diuron to *Selenastrum capricornutum* Printz. DuPont study number AMR-2046-91. ABC laboratory project ID, final report #39335. Unpublished study prepared by ABC Laboratories, Inc. Columbia, MO, submitted to the U. S. Environmental Protection Agency. EPA MRID 42218401.

Cain JR, Cain RK (1983) The effects of selected herbicides on zygospore germination and growth of *Chlamydomonas moewusii* (Chlorophyceae, Volvocales). J Phycology 19:301–305.

Call DJ, Brooke LT, Kent RJ (1983) Toxicity, Bioconcentration and Metabolism of 5 Herbicides in Freshwater Fish. Environmental Research Laboratory-Duluth. U. S. Environmental Protection Agency report, EPA MRID 00141636/TRID 452601029.

Call DJ, Brooke LT, Kent RJ, Knuth ML, Poirier SH, Huot JM, Lima AR (1987) Bromacil and Diuron Herbicides - Toxicity, Uptake, and Elimination in Freshwater Fish. Arch Environ Contam Toxicol 16:607–613.

CARB (2008) California Ambient Air Quality Standards (CAAQS). California Air Resources Board, Sacramento, CA.

CDFG (2010a) State and federally listed endangered and threatened animals of California. California Natural Diversity Database. California Department of Fish and Game, Sacramento, CA. Available from: http://www.dfg.ca.gov/biogeodata/cnddb/pdfs/TEAnimals.pdf.

CDFG (2010b) State and federally listed endangered, threatened, and rare plants of California. California Natural Diversity Database. California Department of Fish and Game, Sacramento, CA. Available from: http://www.dfg.ca.gov/biogeodata/cnddb/pdfs/TEPlants.pdf.

CDWR (1995) Compilation of Sediment and Soil Standards, Criteria, and Guidelines. California Department of Water Resources, State of California, The Resources Agency, Sacramento, CA.

Chesworth JC, Donkin ME, Brown MT (2004) The interactive effects of the antifouling herbicides Irgarol 1051 and Diuron on the seagrass *Zostera marina* (L.). Aquat Toxicol 66:293–305.

Christian FA, Tate TM (1983) Toxicity of Fluometuron and Diuron on the Intermediate Snail Host (*Lymnaea Spp*) of *Fasciola hepatica*. Bull Environ Contam Toxicol 30:628–631.

Crommentuijn T, Kalf DF, Polder MD, Posthumus R, van de Plassche EJ (1997) Maximum permissible concentrations and negligible concentrations for pesticides. RIVM report number 601501002. National Institute of Public Health and the Environment, Bilthoven, The Netherlands.

Crommentuijn T, Sijm D, de Bruijn J, van Leeuwen K, van de Plassche E (2000) Maximum permissible and negligible concentrations for some organic substances and pesticides. J Environ Manag 58:297–312.

Crosby DG, Tucker RK (1966) Toxicity of Aquatic Herbicides to *Daphnia magna*. Science 154:289–291.

Dengler D (2006a) Testing of toxic effects of diuron technical on the blue-green alga *Synechococcus leopoliensis*. Final report. Unpublished study prepared by GAB Biotechnologie GmbH & GAB Analytik GmbH, Germany, sponsored by DuPont de Nemours France S.A. Crop Protection, submitted to the U.S. Environmental Protection Agency. EPA MRID 47936501.

Dengler D (2006b) Testing of toxic effects of diuron technical on the diatom *Navicula pelliculosa*. Final report. Unpublished study prepared by GAB Biotechnologie GmbH & GAB Analytik GmbH, sponsored by DuPont de Nemours France S.A. Crop Protection, submitted to the U.S. Environmental Protection Agency. EPA MRID 47936502.

Devilla RA, Brown MT, Donkin M, Tarran GA, Aiken J, Readman JW (2005) Impact of antifouling booster biocides on single microalgal species and on a natural marine phytoplankton community. Marine Ecology-Progress Series 286:1–12.

Dorigo U, Leboulanger C, Berard A, Bouchez A, Humbert JF, Montuelle B (2007) Lotic biofilm community structure and pesticide tolerance along a contamination gradient in a vineyard area. Aquat Microbial Ecol 50:91–102.

Douglas MT, Handley JW (1988) The algistatic activity of diuron technical. Unpublished study prepared by Huntingdon Research Centre Ltd., Huntingdon, England, sponsored by Du Pont de Nemours (France) S. A., submitted to the U.S. Environmental Protection Agency. EPA MRID 47936503.

Ensminger MP, Starner K, Kelley K (2008) Simazine, diuron, and atrazine detections in California surface waters. California Department of Pesticide Regulation, Sacramento, CA.

Eullaffroy P, Frankart C, Biagianti S (2007) Toxic effect assessment of pollutant mixtures in *Lemna minor* by using polyphasic fluorescence kinetics. Toxicol Environ Chem 89:683–393.

Eullaffroy P, Vernet G (2003) The F684/F735 chlorophyll fluorescence ratio: a potential tool for rapid detection and determination of herbicide phytotoxicity in algae. Water Res 37:1983–1990.

Fernandez-Alba AR, Hernando MD, Piedra L, Chisti Y (2002) Toxicity evaluation of single and mixed antifouling biocides measured with acute toxicity bioassays. Anal Chim Acta 456:303–312.

Ferrell BD (2006) Diuron (DPX-14740) technical: Static, 7-day growth inhibition toxicity test with *Lemna gibba* G3. Laboratory project ID: DuPont 20775. Unpublished study prepared by E.I. du Pont de Nemours and Company Haskell Laboratory for Health and Environmental Sciences, Newark, DE, submitted to the U.S. Environmental Protection Agency. EPA MRID 46996701.

Flum TF, Shannon LJ (1987) The Effects of 3 Related Amides on Microecosystem Stability. Ecotoxicol Environ Saf 13:239–252.

Gatidou G, Thomaidis NS (2007) Evaluation of single and joint toxic effects of two antifouling biocides, their main metabolites and copper using phytoplankton bioassays. Aquat Toxicol 85:184–191.

Geoffroy L, Teisseire H, Couderchet M, Vernet G (2002) Effect of oxyfluorfen and diuron alone and in mixture on antioxidative enzymes of *Scenedesmus obliquus*. Pestic Biochem Physiol 72:178–185.

Hansch C, Leo A, Hoekman D (1995) Exploring QSAR. Hydrophobic, Electronic, and Steric Constants. American Chemical Society, Washington, DC.

Hartgers EM, Aalderink GH, Van Den Brink PJ, Gylstra R, Wiegman JWF, Brock TCM (1998) Ecotoxicological threshold levels of a mixture of herbicides (Atrazine, diuron and metolachlor) in freshwater microcosms. Aquat Ecol 32:135–152.

Hernando MD, Ejerhoon M, Fernandez-Alba AR, Chisti Y (2003) Combined toxicity effects of MTBE and pesticides measured with *Vibrio fischeri* and *Daphnia magna* bioassays. Water Res 37:4091–4098.

Hollister T, Walsh GE (1973) Differential responses of marine phytoplankton to herbicides - oxygen evolution. Bull Environ Contam Toxicol 9:291–295.

Isensee AR (1976) Variability of Aquatic Model Ecosystem-Derived Data. Int J Environ Studies 10:35–41.

IUPAC (2008) IUPAC Agrochemical Information - Diuron. URL ≤http://sitem.herts.ac.uk/aeru/iupac/260.htm≥.

Johnson WW, Finley MT (1980) Handbook of Acute Toxicity of Chemicals to Fish and Aquatic Invertebrates. Resource Publication 137. United States Fish and Wildlife Service, Washington, DC. EPA MRID 40094602.

Killeen S (1997) Development and use of environmental quality standards (EQS) for priority pesticides. Pestic Sci 49:191–195.

Knauer K, Sobek A, Bucheli TD (2007) Reduced toxicity of diuron to the freshwater green alga *Pseudokirchneriella subcapitata* in the presence of black carbon. Aquat Toxicol 83:143–148.

Knauert S, Escher B, Singer H, Hollender J, Knauer K (2008) Mixture toxicity of three photosystem II inhibitors (atrazine, isoproturon, and diuron) toward photosynthesis of freshwater phytoplankton studied in outdoor mesocosms. Environ Sci Technol 42:6424–6430.

Koutsaftis A, Aoyama I (2007) Toxicity of four antifouling biocides and their mixtures on the brine shrimp *Artemia salina*. Sci Total Environ 387:166–174.

Lambert SJ, Thomas KV, Davy AJ (2006) Assessment of the risk posed by the antifouling booster biocides Irgarol 1051 and diuron to freshwater macrophytes. Chemosphere 63:734–743.

Lide DR (ed) (2003) Handbook of Chemistry and Physics. 84th Edition. CRC Press, Boca Raton, FL.

Lydy MJ, Austin KR (2005) Toxicity assessment of pesticide mixtures typical of the Sacramento-San Joaquin Delta using *Chironomus tentans*. Arch Environ Contam Toxicol 48:49–55.

Ma J, Liang W, Xu L, Wang S, Wei Y, Lu J (2001) Acute toxicity of 33 herbicides to the green alga *Chlorella pyrenoidosa*. Bull Environ Contam Toxicol 66:536–541.

Ma J (2002) Differential sensitivity to 30 herbicides among populations of two green algae *Scenedesmus obliquus* and *Chlorella pyrenoidosa*. Bull Environ Contam Toxicol 68:275–281.

Ma J, Lin F, Wang S, Xu L (2003) Toxicity of 21 herbicides to the green alga *Scenedesmus quadricauda*. Bull Environ Contam Toxicol 71:594–601.

Ma JY, Wang SF, Wang PW, Ma LJ, Chen XL, Xu RF (2006) Toxicity assessment of 40 herbicides to the green alga *Raphidocelis subcapitata*. Ecotoxicol Environ Saf 63:456–462.

Ma JY, Xu LG, Wang SF, Zheng RQ, Jin SH, Huang SQ, Huang YJ (2002) Toxicity of 40 herbicides to the green alga *Chlorella vulgaris*. Ecotoxicol Environ Saf 51:128–132.

Macek KJ, Hutchins C, Cope OB (1969) Effects of Temperature on Susceptibility of Bluegills and Rainbow Trout to Selected Pesticides. Bull Environ Contam Toxicol 4:174–183.

Mackay D, Shiu WY, Ma KC, Lee SC (2006) Handbook of Physical-Chemical Properties and Environmental Fate for Organic Chemicals. 2nd edn. CRC Press, Boca Raton, FL.

Manzo S, Buono S, Cremisini C (2008) Predictability of copper, irgarol, and diuron combined effects on sea urchin *Paracentrotus lividus*. Arch Environ Contam Toxicol 54:57–68.

Maule A, Wright SJL (1984) Herbicide effects on the population-growth of some green-algae and cyanobacteria. J Appl Bacteriol 57:369–379.

Menconi M, Beckman J (1996) Hazard assessment of the insecticide methomyl to aquatic organisms in the San Joaquin river system. Administrative report 96–6. California Department of Fish and Game, Rancho Cordova, CA.

Molander S, Blanck H (1992) Detection of Pollution-Induced Community Tolerance (Pict) in Marine Periphyton Communities Established under Diuron Exposure. Aquat Toxicol 22:129–144.

Molander S, Dahl B, Blanck H, Jonsson J, Sjostrom M (1992) Combined Effects of Tri-Normal-Butyl Tin (Tbt) and Diuron on Marine Periphyton Communities Detected as Pollution-Induced Community Tolerance. Arch Environ Contam Toxicol 22:419–427.

Mount DR, Ankley GT, Brix KV, Clements WH, Dixon DG, Fairbrother A, Hickey CW, Lanno RP, Lee CM, Munns WR, Ringer RK, Staveley JP, Wood CM, Erickson RJ, Hodson PV (2003) Effects assessment: Introduction. In: *Reevaluation of the State of the Science for Water-Quality Criteria Development*, Reiley MC, Stubblefield WA, Adams WJ, Di Toro DM, Hodson PV, Erickson RJ, Keating FJ Jr, eds., SETAC Press, Pensacola, FL.

Nebeker AV, Schuytema GS (1998) Chronic effects of the herbicide diuron on freshwater cladocerans, amphipods, midges, minnows, worms, and snails. Arch Environ Contam Toxicol 35:441–446.

NOAA (1999) Sediment Quality Guidelines Developed for the National Status and Trends Program. National Oceanographic and Atmospheric Agency Office of Response and Restoration, Department of Commerce.

Okamura H, Nishida T, Ono Y, Shim WJ (2003) Phytotoxic effects of antifouling compounds on nontarget plant species. Bull Environ Contam Toxicol 71:881–886.

Okamura H, Watanabe T, Aoyama I, Hasobe M (2002) Toxicity evaluation of new antifouling compounds using suspension-cultured fish cells. Chemosphere 46:945–951.

Palumbo AJ, TenBrook PL, Fojut TL, Faria IR, Tjeerdema RS (2012) Aquatic life water quality criteria derived via the UC Davis method: I. Organophosphate insecticides. Rev Environ Contam Toxicol 216:1–49.

Perschbacher PW, Ludwig GM (2004) Effects of diuron and other aerially applied cotton herbicides and defoliants on the plankton communities of aquaculture ponds. Aquaculture 233:197–203.

Pesce S, Fajon C, Bardot C, Bonnemoy F, Portelli C, Bohatier J (2006) Effects of the phenylurea herbicide diuron on natural riverine microbial communities in an experimental study. Aquat Toxicol 78:303–314.

Podola B, Melkonian M (2005) Selective real-time herbicide monitoring by an array chip biosensor employing diverse microalgae. J Appl Phycol 17:261–271.

Raimondo S, Vivian DN, Barron MG (2010) Web-based Interspecies Correlation Estimation (Web-ICE) for Acute Toxicity: User Manual. Version 3.1. Office of Research and Development, U.S. Environmental Protection Agency, Gulf Breeze, FL. EPA/600/R-10/004.

Sanders HO (1969) 25. Toxicity of Pesticides to the Crustacean *Gammarus lacustris*. Bureau of Sport Fisheries and Wildlife. United States Department of the Interior Fish and Wildlife Service, Washington, DC.

Sanders HO (1970) Toxicities of some herbicides to 6 species of freshwater crustaceans. J Water Pollut Cont Fed 42:1544–1550.

Sanders HO, Cope OB (1968) Relative Toxicities of Several Pesticides to Naiads of 3 Species of Stoneflies. Limnol Oceanogr 13:112–117.

Sangster Research Laboratories (2008) LOGKOW A databank of evaluated octanol-water partition coefficients (Log P). URL <http://logkow.cisti.nrc.ca/logkow/index.jsp>.

Schafer H, Hettler H, Fritsche U, Pitzen G, Roderer G, Wenzel A (1994) Biotests using unicellular algae and ciliates for predicting long-term effects of toxicants. Ecotoxicol Environ Saf 27:64–81.

Schrader KK, de Regt MQ, Tidwell PD, Tucker CS, Duke SO (1998) Compounds with selective toxicity towards the off-flavor metabolite-producing cyanobacterium *Oscillatoria cf. chalybea*. Aquaculture 163:85–99.

Schuytema GS, Nebeker AV (1998) Comparative toxicity of diuron on survival and growth of Pacific treefrog, bullfrog, red-legged frog, and African clawed frog embryos and tadpoles. Arch Environ Contam Toxicol 34:370–376.

Siepmann S, Jones MR (1998) Hazard assessment of the insecticide carbaryl to aquatic organisms in the Sacramento-San Joaquin River system. Administrative report 98–1. California Department of Fish and Game, Office of Spill Prevention and Response, Rancho Cordova, CA.

Sumpono, Perotti P, Belan A, Forestier C, Lavedrine B, Bohatier J (2003) Effect of Diuron on aquatic bacteria in laboratory-scale wastewater treatment ponds with special reference to *Aeromonas* species studied by colony hybridization. Chemosphere 50:445–455.

Teisseire H, Couderchet M, Vernet G (1999) Phytotoxicity of diuron alone and in combination with copper or folpet on duckweed (*Lemna minor*). Environ Pollut 106:39–45.

TenBrook PL, Palumbo AJ, Fojut TL, Hann P, Karkoski J, Tjeerdema RS (2010) The University of California-Davis Methodology for deriving aquatic life pesticide water quality criteria. Rev Environ Contam Toxicol 209:1–155.

Tlili A, Dorigo U, Montuelle B, Margoum C, Carluer N, Gouy V, Bouchez A, Berard A (2008) Responses of chronically contaminated biofilms to short pulses of diuron – An experimental study simulating flooding events in a small river. Aquat Toxicol 87:252–263.

Tomlin C (2003) The Pesticide Manual, A World Compendium. 13th edition. The British Crop Protection Council, Alton, Hampshire, UK.

Tooby TE, Lucey J, Stott B (1980) The tolerance of grass carp, *Ctenopharyngodon idella* val to aquatic herbicides. J Fish Biol 16:591–597.

Ukeles R (1962) Growth of pure cultures of marine phytoplankton in presence of toxicants. Appl Microbiol 10:532–537.

USEPA (1985) Guidelines for deriving numerical national water quality criteria for the protection of aquatic organisms and their uses, PB-85-227049. United States Environmental Protection Agency, National Technical Information Service, Springfield, VA.

USEPA (1996) Algal Toxicity, Tiers I and II, Ecological Effects Test Guidelines, OPPTS 850.5400, EPA 712/C/96/164. United States Environmental Protection Agency, Washington, DC.

USEPA (2003) Reregistration Eligibility Decision (RED) for Diuron. United States Environmental Protection Agency, Office of Prevention, Pesticides, and Toxic Substances, Washington, DC.

USEPA (2007) Diuron, Pesticide Tolerance. Federal Register, Docket # EPA-HQ-OPP-2006-0559, 72, 32533–32540.

USFDA (2000) Industry Activities Staff Booklet. URL <http://www.cfsan.fda.gov/~lrd/fdaact.html>.

USFWS (2010) Species Reports. Endangered Species Program. U.S. Fish and Wildlife Service. Available from: http://www.fws.gov/endangered/; http://ecos.fws.gov/tess_public/pub/listedAnimals.jsp; http://ecos.fws.gov/tess_public/pub/listedPlants.jsp.

Walker CR (1965) Diuron, fenuron, monuron, neburon, and TCA mixtures as aquatic herbicides in fish habitats. Weeds 13:297–301.

Walsh GE (1972) Effects of Herbicides on Photosynthesis and Growth of Marine Unicellular Algae. Water Hyacinth J 10:45–48.

Walsh GE, Grow TE (1971) Depression of Carbohydrate in Marine Algae by Urea Herbicides. Weed Sci 19:568–570.

Ward T, Boeri R (1991) Acute Flow-through Mollusc Shell Deposition Test with DPX-14740-166 (Diuron). Haskell laboratory outside report No. MR-4581-911. Unpublished study prepared by EnviroSystems Division Resource Analysts, Inc., Hampton, NH, sponsored by E.I. du Pont de Nemours and Company, Newark, DE, submitted to the U. S. Environmental Protection Agency. EPA MRID 42217201.

Ward T, Boeri R (1992a) Early life stage toxicity of DPX-14740-166 (Diuron) to Sheepshead minnow, *Cyprinodon variegatus*. Haskell laboratory outside report No. 866-91. Unpublished study prepared by EnviroSystems Division Resource Analysts, Inc., Hampton, NH, sponsored by E.I. du Pont de Nemours and Company, Newark, DE, submitted to the U. S. Environmental Protection Agency. EPA MRID 42312901.

Ward T, Boeri R (1992b) Life-cycle Toxicity of DPX-14740-166 (Diuron) to the Mysid, *Mysidopsis bahia*. Haskell laboratory outside report No. 203-92. Unpublished study prepared by EnviroSystems Division Resource Analysts, Inc., Hampton, NH, sponsored by E.I. du Pont de Nemours and Company, Newark, DE, submitted to the U. S. Environmental Protection Agency. EPA MRID 42500601.

Zimba PV, Tucker CS, Mischke CC, Grimm CC (2002) Short-term effect of diuron on catfish pond ecology. North Am J Aquacult 64:16–23.

Index

R.S. Tjeerdema (ed.), *Aquatic Life Water Quality Criteria for Selected Pesticides*, Reviews of Environmental Contamination and Toxicology 216, DOI 10.1007/978-1-4614-2260-0, © Springer Science+Business Media, LLC 2012